Bin Han

Characterization and Emulation of Low-Voltage Power Line Channels for Narrowband and Broadband Communication

Forschungsberichte aus der Industriellen Informationstechnik
Band 14

Institut für Industrielle Informationstechnik
Karlsruher Institut für Technologie
Hrsg. Prof. Dr.-Ing. Fernando Puente León
Prof. Dr.-Ing. habil. Klaus Dostert

Eine Übersicht aller bisher in dieser Schriftenreihe erschienenen Bände finden Sie am Ende des Buchs.

Characterization and Emulation of Low-Voltage Power Line Channels for Narrowband and Broadband Communication

by
Bin Han

KIT Scientific Publishing

Dissertation, Karlsruher Institut für Technologie
KIT-Fakultät für Elektrotechnik und Informationstechnik

Tag der mündlichen Prüfung: 31. Mai 2016
Referenten: Prof. Dr.-Ing. habil. Klaus Dostert
Prof. Dr.-Ing. Dr. h. c. Jürgen Becker

Impressum

KIT Scientific Publishing

Karlsruher Institut für Technologie (KIT)
KIT Scientific Publishing
Straße am Forum 2
D-76131 Karlsruhe

KIT Scientific Publishing is a registered trademark
of Karlsruhe Institute of Technology.
Reprint using the book cover is not allowed.

www.ksp.kit.edu

Print on Demand 2017 – Gedruckt auf FSC-zertifiziertem Papier

ISSN 2190-6629
ISBN 978-3-7315-0654-6
DOI 10.5445/KSP/1000068355

Dedicated to Shuoren Li

Acknowledgements

Firstly, I would like to express my sincere gratitude to my advisor Prof. Dr.-Ing. habil. Klaus Dostert, for his continuous support of my Ph.D. study and related research, for his patience, motivation, and immense knowledge. His guidance helped me in all the time of research and writing of this thesis. I could not have imagined having a better advisor and mentor for my Ph.D. study.

Besides, I would also like to thank my second advisor, Prof. Dr.-Ing. Dr. h.c. Jürgen Becker, for his insightful comments, for his encouragement and for the hard questions which incented me to widen my research from various perspectives.

My sincere thanks also goes to Prof. Dr.-Ing. Fernando Puente and all my colleagues at Institute of Industrial Information Technology (IIIT), for their continuous support and for providing me with a pleasant and friendly working atmosphere. It would not be possible to conduct the research without their help.

Particularly I appreciate the selfless help from Dr.-Ing. Wenqing Liu, M.Sc. Nico Otterbach, M.Sc. Vlad Stoica, M.Sc. Cornelius Kaiser and China Electric Power Research Institute (CEPRI), which has been greatly assisting me with my research. I would like to express my thanks to Dr.-Ing. Matthias Michelsburg, Dipl.-Ing. Andre Suchaneck, and M.Sc. Thomas Nürnberg, for their great contributions to the LaTeX template of the thesis. I also owe much gratitude to Dipl.-Ing. Damián Ezequiel Alonso, for his valuable suggestions for improving the thesis.

Last but not least, I would like to express my very profound gratitude to my family: my parents and my wife Jiao. Their permanent spiritual support and meticulous care have always been the emotional anchor to me throughout the completion of this thesis and my life in general.

Contents

Abstract

Realizing smart grid and smart home applications calls for reliable and cost-efficient communication systems. This demand has raised the recent interest in power line communication (PLC) technologies, and has driven a broad set of deep surveys in low-voltage (LV) power line channels. Generally, two main challenges are to overcome in this field: the channel characterization and the channel emulation. The first one helps to understand and describe the complex behavior of the channels, so that reliable PLC systems can be pertinently developed. The latter one supports a reproducible, flexible and rapid PLC system evaluation on the physical layer. This thesis proposes a set of novel approaches, to characterize and to emulate LV power line channels in the frequency range from 0.15 to 10 MHz, which closes gaps between the traditional narrowband (up to 500 kHz) and broadband (above 1.8 MHz) ranges. Measuring frameworks are designed to obtain the cyclostationary channel properties, including the channel transfer function (CTF) and the interference scenario. Existing models of CTF and noise are reviewed and extended to describe the channel behavior. According to the extended models, adaptive and robust estimators are implemented to parameterize the channel. Supported by the estimated parameters, an advanced FPGA-based high-speed channel emulator is implemented to reconstruct LV power line channels in real-time. Based on the proposed emulator, an LV PLC testbed is developed to support system evaluation for both narrowband and broadband applications.

Zusammenfassung

Die Zielsetzung dieser Arbeit besteht darin, das komplexe Verhalten von Niederspannungsstromnetzen als Datenübertragungskanäle, - d.h. ihre Kanalübertragungsfunktionen (CTF) und die Störszenarien - im Frequenzbereich zwischen 0,15 und 10 MHz mit Berücksichtigung der Zyklostationarität zu charakterisieren und zu emulieren. Das Phänomen der Zyklostationarität wird umfassend untersucht und die Auswirkungen auf die Powerline-Kommunikation (PLC) werden diskutiert. Ein verteiltes Messsystem zur Messung der Übertragungsfunktion wird vorgestellt und zwei unterschiedliche Anregungssignaltypen werden unter Annahme verschiedener Störbedingungen simulativ verglichen. Eine Methode von Clustering wird entwickelt, um gemessene zeitvariante Übertragungsfunktionen mit einer Sammlung von Mustern zeitinvarianter Übertragungsfunktionen mittles periodischer Umschaltung adaptiv zu approximieren. Verfahren zur Erfassung komplexer Störsignalkombinationen werden entworfen und damit gewonnene Messergebnisse werden umfassend dargestellt. Basierend auf Vorkenntnissen über die Klassifikation von Störungen in Stromnetzen sowie bekannten Störmodellen, werden neuartige Verfahren entwickelt. Mit diesen Verfahren werden aus Störsignalmessungen voneinander unabhängige Störer extrahiert und die charakteristische Parameter jedes Störers werden abgeschätzt. Ein Kanalemulator wird auf FPGA-Basis entwickelt, um alle aus den Messungen und deren Analyse gewonnenen Erkenntnisse echtzeitfähig nachzubilden. Auf der Basis des Emulators wird schließlich eine vollständige Niederspannungs-PLC-Testumgebung implementiert, mit der Powerline-Kommunikationssysteme im Frequenzbereich 0,15 - 10 MHz flexibel, objektiv und jederzeit reproduzierbar evaluiert werden können.

Nomenclature

Common abbreviations

Abbreviation	Description
e. g.	exemplī grātiā (for example)
i. e.	id est (it is)
etc.	et cetera (and the rest)
2PN	two-port network
ACF	autocorrelation function
AFE	analog front-end
AIN	asynchronous impulsive noise
AM	amplitude modulated
AMR	automated meter reading
APIN	aperiodic impulsive noise
AR	autoregressive
ARIB	*Association of Radio Industries and Businesses*
ARMA	autoregressive moving-average
ASK	amplitude shift keying
AWGN	additive white Gaussian noise
BB	broadband
bps	bit per second
BPSK	binary phase shift keying
CBGN	colored background noise
CCF	cyclic coherence function
CDF	cumulative distibution function
CENELEC	*Comité Européen de Normalisation Électrotechnique*
CM	common mode
CS2	second-order cyclostationary
CSD	cyclic spectral density

Abbreviation	Description
DBPSK	differential binary phase shift keying
DCSK	differential code shift keying
DFT	discrete Fourier transform
DLL	data link layer
DM	differential mode
DQPSK	differential qudrature phase shift keying
DSM	demand side management
DUT	device under test
EMD	empirical mode decomposition
EM	electro-magnetic
FM	frequency modulated
FPGA	field programmable gate array
HDR	high data rate
HF	high frequency
HHT	Hilbert-Huang transform
HV	high voltage
IMF	intrinsic mode funtion
ITU-T	*International Telecommunications Union - Telecommunication Standardization Sector*
ITU	*International Telecommunications Union*
LDR	low data rate
LF	low frequency
LFSR	linear feedback shift register
LISN	line impedance stabilization network
LPTV	linear periodically time-varying
LSTV	linear switching time-varying
LSW	locally stationary wavelet
LTI	linear time-invariant
LUT	look-up table
LV	low voltage
M-sequence	maximal length binary sequence
MAC	media access control
MA	moving average
MCR	multiple cyclic regression
MF	medium frequency
MIMO	multi-input multi-output

Abbreviation	Description
MSE	mean square error
MV	medium voltage
NB	narrowband
NBN	narrowband noise
OLA	overlap-add
OLS	overlap-save
Op-Amp	operational amplifier
PACF	partial autocorrelation function
PC	personal computer
PDF	probability density function
PE	protective earth
PHY	physical layer
PINAS	periodic impulsive noise asynchronous to the mains voltage
PIN	periodic impulsive noise
PINS	periodic impulsive noise synchronous to the mains voltage
PLC	power line communication
PLS	physical signaling
PN	pseudo noise
PSD	power spectral density
PSK	phase shift keying
QAM	qudrature amplitude modulation
QPSK	qudrature phase shift keying
RAM	random-access memory
RLS	recursive least squares
SBL	sparse Bayesian learning
SISO	single-input single-output
SLF	super low frequency
SNR	signal-to-noise ratio
SQNR	signal-to-quantization noise
STFT	short-time Fourier transform
TL	transmission line
UDRN	uniformly distributed random numbers
UHF	ultra high frequency
ULF	ultra low frequency

Abbreviation	Description
UNB	ultra narrow band
UPS	uninterruptible power supply
US FCC	*United States Federal Communications Commission*
USRP	universal software radio peripheral
VGA	variable gain amplifier
VLF	very low frequency
VNA	vector network analyzer
ZCD	zero-crossing detector

Letters

Latin Letters

Symbol	Description
B	Bandwidth
C_{yy}	Covariance
f	Frequency
F_y	Cumulative distribution function
f_y	Probability density function
H	Transfer function
i	Current
$\mathbb{R}$	Real numbers
R_{yy}	Autocorrelation function
t	Time
u	Voltage
$\mathbb{Z}$	Integers
Z	Impedance

Greek Letters

Symbol	Description
δ	Dirac function
φ	Phase offset
τ	Delay

Superscripts

Index	Description
$(\bullet)^{\mathrm{T}}$	Transpose

Subscripts

Index	Description
$(\bullet)_{\mathrm{Cal}}$	Calibration
$(\bullet)_{\mathrm{Mea}}$	Measurement
$(\bullet)_{\mathrm{Ref}}$	Reference
$(\bullet)_{\mathrm{R}}$	Receiver
$(\bullet)_{\mathrm{T}}$	Transmitter
$(\bullet)_{\mathrm{Unc}}$	Uncalibrated

Mathematical operators

Operator	Description
$\simeq$	Approximation
$\in$	Belong to
$*$	Convolution
$\triangleq$	Definition
$\mathrm{E}\{\bullet\}$	Expectation
$\lfloor\bullet\rfloor$	Floor
$\forall$	For all
$\cdot$	Multiplication
$\angle\{\bullet\}$	Phase
$\cup$	Union

1 Introduction

1.1 Background and Motivation

Since the first establishment in the early years of the 20th century, electrical grids have been pervasively erected over urban, suburb and rural areas all over the planet, and become an indispensable part of the modern world. After one century's rapid development of technologies, demand for new functionalities of electrical grids have been called, such as automated meter reading (AMR), load control, demand side management (DSM) and remote diagnostics. The concept of 'smart grid' was therefore proposed, which represents an intelligent power network, based on the existing electrical grids and reliable communication technologies [30]. Due to its unique advantages of ubiquitous infrastructure, easy access and low additional building cost, power line communication (PLC) has been usually considered as a competitive solution for the communication backbone of smart grids, and has thus raised a dramatical amount of interest in the past decade.

However, the existing power grids were originally designed to deliver no data but only energy. As telecommunication medium they generally exhibit harsh channel conditions, such as strong attenuation, high noise power, complex interference scenarios, frequency selectivity and cyclostationarity. This has been challenging the deployment of PLC regarding standard establishment, device development and system evaluation. In the past years, great efforts have been made to investigate power line channels in narrowband (3 to 500 kHz) and broadband (above 1.8 MHz). In each band the dominant channel characteristics have been deeply studied, leading to the establishment of current PLC standards and specifications (see Sec. 2.4). However, when attempting to apply PLC in a new frequency range across narrowband and broadband, e.g. between 0.15 to 10 MHz [54], the existing methods of power line channel characterization and PLC device evaluation are no longer feasible. New approaches are therefore demanded.

1.2 State of the Art

In the researching field of PLC, channel characterization has been a popular subject since the 1990s, especially focusing on the low-voltage (LV) level, including the access domain and the indoor domain. A huge amount of measurements and a variety of models have been proposed to describe the power line channel behaviors with respect to transfer function [2–5, 13, 14, 24, 27–29, 37, 42, 48, 58, 61, 72, 87, 90, 96, 99], access impedance [14, 42, 77, 82, 90] and interference scenario [11, 14, 16, 17, 24, 37, 42, 44–46, 58, 65, 68, 71, 72, 98, 100]. In the recent years, multi-input multi-output (MIMO) measurements and models have been more and more reported [39–41, 76, 79, 81, 84, 85, 89]. Generally, it can be concluded that the power line channels suffer from time-varying frequency-selective fading and adverse noise scenarios [8].

These harsh characteristics of power line channels have set strong challenges not only for the development of PLC devices, but also for the evaluation of them, because a stable test environment can be neither set in a real grid, nor easily reproduced with ordinary laboratory equipment. To provide a flexible solution of this problem, generations of power line channel emulators based on field programmable gate arrays (FPGAs) have been designed for narrowband [6, 55, 57, 58] and broadband [15, 37, 92–94] applications.

1.3 Goals and Challenges

Focusing on the frequency range 0.15 to 10 MHz, this work has two goals:

- To develop a set of systematical approaches for LV power line channel characterization, including the measurement and the parameter estimation of the channel, with respect to the transfer function and the noise scenarios.

- To develop an advanced FPGA-based power line channel emulator, and a corresponding PLC testbed. Unidirectional transfer functions and noise scenarios shall be emulated with consideration of cyclostationarity, to provide a flexible evaluation of PLC devices on the physical layer (PHY).

In order to achieve the listed goals, several challenges are to overcome:

Selecting the Channel Sounding Signal

Due to the large spacial dimension of power line networks, their transfer functions are usually measured in a distributed approach, i.e. a channel sounder and two externally synchronized receivers are used, the channel sounder stimulates the channel under measurement, while the two receivers capture the sounding signal and the the channel output, respectively. The channel transfer function is then estimated by correlating the sounding signal and the channel output. Generally, two kinds of sounding signals have been demonstrated to be available for measuring power line channels: wide-band time sequences such as pseudo noise sequences [97], and sweeping tone(s) [55]. In both cases of sounding signal, the accuracy of measurement depends on the signal-to-noise ratio (SNR), and is therefore limited by the maximal output power of the channel sounder. When considering the time variation of the transfer function, the total duration of measurement critically impacts the precision of results as well. Till now, there is no reported study comparing the performance of the two sounding methods under the same constrains of power and time. To achieve the best accuracy of measurement, the sounding signal must be carefully selected, based on performance evaluation.

Segmenting Time-Varying Channel Status

It is a common approximation to model a time-varying process as a series of time-invariant processes. This method has also been widely applied in power line channel characterization, especially in the modeling of channel transfer functions and background noises. However, the fluctuation of these features is usually non-linear in time, hence, a simple uniform time segmentation cannot promise a satisfying estimation; an advanced method is therefore required.

Detecting and Extracting Interferences

Noise at power line channels is a complex mixture of multiple interferences, which are completely distinct from each other in power spectral distribution and time behavior. Mature models have been developed in the past decades for each class of interference, but the estimation of model parameters can be difficult with coexistence of interferences of other types. To reduce the effort of manual analysis, systematic and automatic methods are needed, which detect and extract the interferences of different classes separately.

Estimating Interference Parameters

Parameter estimation plays the key role of connecting the analytic channel characterization and the synthetic channel emulation. Compared to the transfer function, the interference parameters are more challenging to estimate, due to the sophisticated noise scenarios. For each interference parameter, a robust estimator shall be designed according to the properly selected model.

Emulating Cyclostationary Channels

Although copious implementations of power line channel emulators have been reported, most of them only emulate time-invariant channels. To realize channel emulation regarding the cyclic behavior, especially the cyclostationary transfer functions, an advanced architecture with complex timing control and mains phase acquisition shall be designed.

1.4 Structure of the Thesis

The thesis is organized as follows: Ch. 2 gives a general introduction to PLC, including the topologies of power line networks and basic principles of deploying power lines as data transmission channels. Cyclostationary phenomena in power line channels are discussed, currently available PLC standards and specifications in different bands are reviewed. In Ch. 3 the characterization of LV power line channels is studied in depth. Acquisition and modeling of channel transfer functions are discussed, focusing on the selection of sounding signals, and the time-varying models of transfer functions. Afterwards, it is proposed how to measure and analyze the noise at power lines. Noises of different classes are sequentially discussed. For each noise class, the existing models are reviewed and the best fitting one is selected. Algorithms for interference detection, extraction and parameter estimation are then designed according to the selected model. Ch. 4 proposes the design of a new PLC testbed, based on a power line channel emulator, and the FPGA implementation of the emulator. Ch. 5 closes the thesis with conclusions and discussions of future work.

2 Narrowband and Broadband Powerline Communications

This chapter introduces power line channels and PLC systems. First the power line network topologies are studied, focusing on the low-voltage level. Then the data transmission over power lines is introduced, followed by two approaches to characterize the channels. As a special feature of power line channels and the one causing main difficulties for PLC, cyclostationarity is discussed. At last an overview of the existing PLC technologies and standards in different frequency bands is given.

2.1 Topologies of Low-Voltage Power Line Networks

To begin the studies of power line channels, the topology of power line networks is first investigated. Power supply networks are usually divided into three levels: high voltage (HV, 110 to 380 kV), medium voltage (MV, 10 to 30 kV) and low voltage (LV, $\leq 0.4\,\mathrm{kV}$), as illustrated in Fig. 2.1. In principle, all three levels are available for communications [9][26][50]. Due to its extended distribution, easy access and low cost for signal coupling, the LV level has attracted the most interest in the PLC field. In this work, this level will be in our focus.

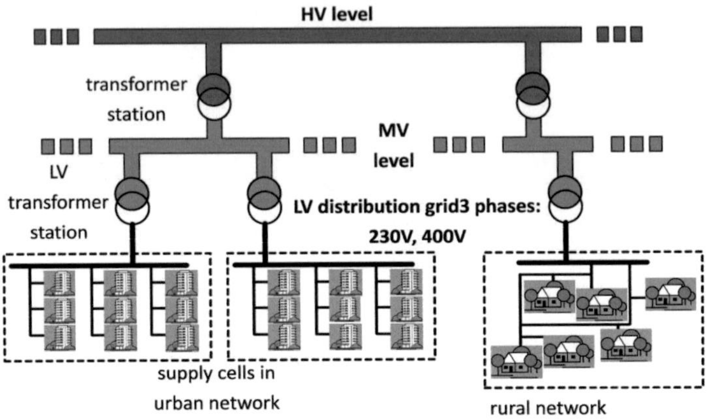

Figure 2.1 Typical electricity supply network topology in Europe [58]

Although the HV and MV supply structures are very similar all over the world, the LV distribution grids can differ significantly from one area to another. For example, the 50 Hz three-phase system is deployed in Europe, Africa and most Asian countries, such as China and India, with residential voltage of either 230 V or 220 V. In most American countries, such as the USA, Canada and Brazil, and some Asian countries, such as Japan, the 60 Hz split-phase system is deployed, with different residential voltages ranging from 100 V to 127 V .

In the view of PLC, a LV network can be further geographically divided into two parts: the access domain and the indoor network. The access domain is usually defined as the LV network between LV transformer stations and building connections. The indoor network, as its name prompts, includes all connections inside buildings. In the following part of this section, different structures of LV distribution grids are compared in both domains.

2.1.1 Access Domain

As shown in Fig. 2.1, in a three-phase system, each LV transformer station supplies a so-called supply cell, which may include up to 350 households in up to 10 branches. This size is often reduced in rural networks. A branch may extend up to 1 km, when the loads are mainly one family houses, for example. Subgroups of several households are always connected with shorter cables [9]. For the connection it is common to use four-wire supply cables, which include three live wires for three phases and one neutral wire, as shown in Fig. 2.2. In some areas, such as China, five-wire cables are also used, which supply an extra wire for the protective earth (PE, or ground). In this case, the outlets in households are connected to the PE wire instead of being locally grounded.

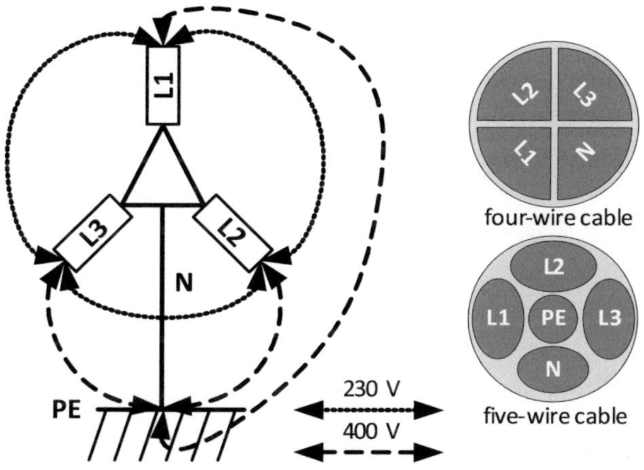

Figure 2.2 Details of 230/400 V three-phase supply and the cables: L, N and PE stand for live, neutral and protective earth, respectively.

In split-phase (also known as single-phase) systems, as shown in Fig. 2.3, each cable entering a household includes three wires: two hot ones and a neutral. It is worth noting that the two hot wires have 180° phase difference, not 90° as in two-phase systems, which were used in the USA in the early 20th century. Compared with the three-phase systems in Europe, the split-phase systems in the USA and Japan usually have small supply cells with less households per transformer and shorter cables, up to about 100 m [26].

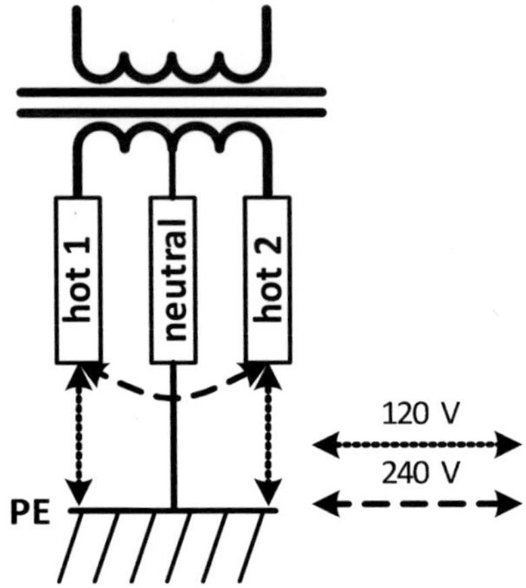

Figure 2.3 Details of 120/240 V split-phase supply

2.1.2 Indoor Domain

Generally, a cable going into a building first passes the main breaker and the meter to reach the user panel, then connects to the wall outlets to distribute the electricity, as shown in Fig. 2.4. Star and tree topologies are standard in almost all countries, except for the UK, where a ring topology is used. With the star or tree configuration, the cables extend into many branches to reach outlets in every wall. With the ring configuration, a single cable runs all the way around a part of a house, sometimes even covering the whole house, especially in old buildings.

The wall outlets can be either grounded (2-wire) or ungrounded (3-wire). In split-phase systems, as depicted in Fig. 2.5, an outlet can be connected between one hot wire and the neutral wire, or between two hot wires, so that 120 V or 240 V voltage can be obtained. In three-phase systems, each outlet is connected between a live wire and the neutral wire. Within old buildings also 5-wire outlets with all three live wires can be found. Different rooms in the same household may be connected to different phases.

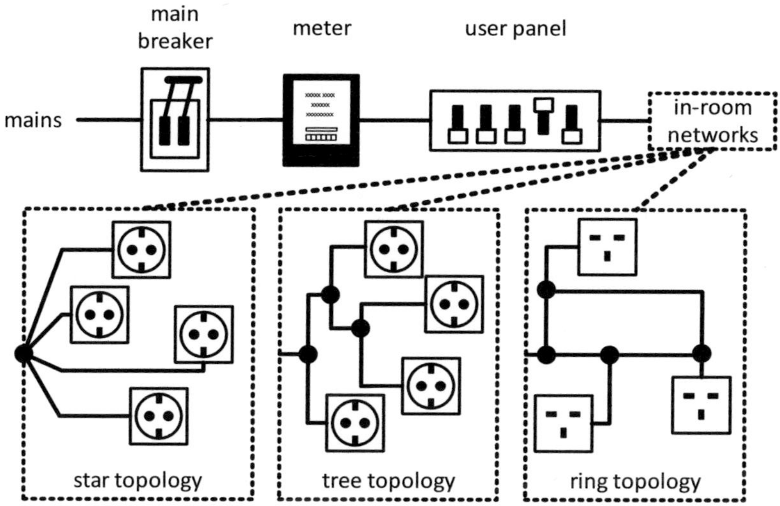

Figure 2.4 Typical topologies of indoor networks

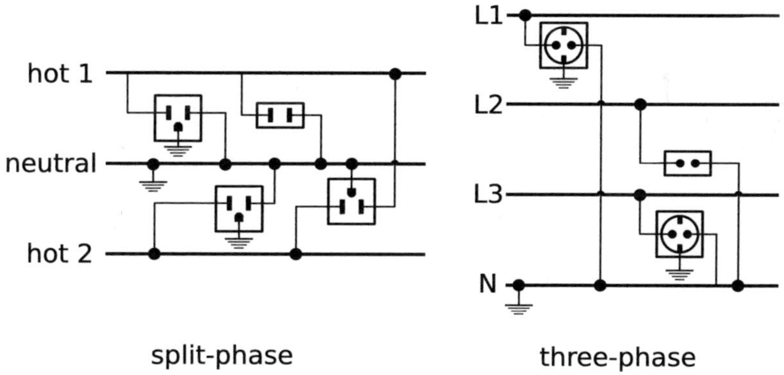

Figure 2.5 Typical indoor wiring systems

2.2 Data Transmission over Power Lines

Although originally designed only for energy delivery, power supply networks have been additionally used for data transmission since a long time. The initial attempt to transmit voice messages over HV networks dates back to the 1920s, motivated by the requirement of a fast bidirectional message flow for maintaining the function of extended HV networks. No telephone network could fully cover all the operating areas at that time, so the power supply networks became an attractive alternative for communication [20]. Today, despite the wide coverage of wired and wireless telephone networks, power line networks still possess an unique advantage of ubiquitous infrastructure as a data medium, especially on the LV level. This section shortly discusses the physical model of data transmission over LV power lines, and the structure of typical PLC devices, then introduces two approaches to characterize power line channels: bottom-up and top-down. Eventually it is shown that any structural model built through the bottom-up approach can be converted into a top-down behavioral model.

2.2.1 Transmission in Common Mode and Differential Mode

As introduced in Sec. 2.1, in LV power line networks, every hot/live wire runs in parallel with a neutral wire, so that a transmission line (TL) in the form of parallel wires is constructed. In addition, each hot/neutral wire together with the PE constructs another TL. Such a TL with a hot wire has usually significantly different characteristics in comparison with the TL with the corresponding neutral wire, because the neutral wire is connected to ground at the LV transformer (and sometimes also in households). Thus, signals can be transmitted over power lines in two different modes: the common mode (CM) and the differential mode (DM), as shown in Fig. 2.6.

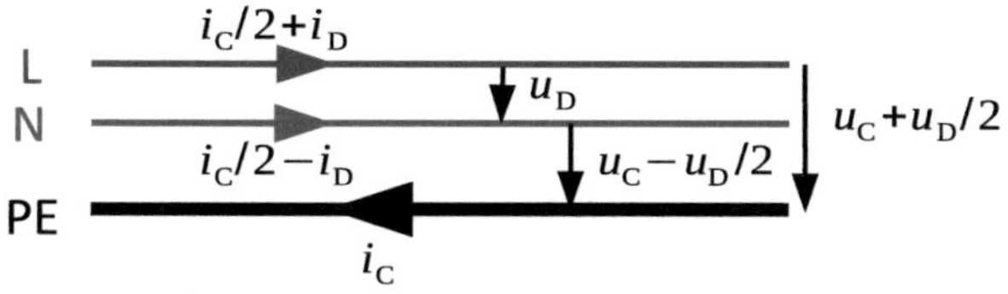

Figure 2.6 Signal can be transmitted over power lines in common and differential modes. u_C and u_D represent the voltage signal in common and differential modes, respectively, i_C and i_D are the corresponding currents [101].

According to studies, CM signals suffer from significantly lower attenuation than DM signals [80]. On the other hand, some kinds of noise are remarkably stronger when received in CM [101]. Traditionally, most single-input single-output (SISO) LV PLC devices transmit/receive signals between a live wire and the corresponding neutral wire, aiming to use only the differential mode. The common mode, in contrast, has attracted interests mainly for its value in MIMO PLC systems, which are not discussed in this thesis. However, the CM and DM components can be partially converted into each other, when the load is imbalanced or asymmetry is present between the two lines. This phenomenon is known as the mode conversion or the mode coupling [26].

Considering about the network topologies introduced in Sec. 2.1, there are three different kinds of communication links in a LV network: links in the same branch/ring, links in different branches on the same phase, and links on different phases. In the last case, communication signals are coupled from one phase to another at the LV transformer. This cross-phase transmission can lead to an attenuation of tens of dBs. In all cases, the wave can be reflected at the ends of branches and/or echoed in loops. Two simple examples are illustrated in Fig. 2.7. In the tree-shaped network shown in Fig. 2.7(a), if the impedance is not matched at nodes A and B (which is the common case in reality), the wave can be echoed between the nodes, before it is finally propagating to the receiver R. In the ring-shaped network shown in Fig. 2.7(b), the wave can echo through the loop along the nodes A-B-C-D or A-D-C-B an arbitrary number of times, before it is finally received by the receiver R, even if the impedance is matched at all nodes. In both cases, depending on times of echoes, multiple propagation paths are present. Thus, the channel always exhibits multi-path properties, i.e. notches and phase shifting. The notches, according to [26], can become much more significant by bonding the neutral wire to ground in the building. This phenomenon, once again, reveals the fact that multi-path fading is weaker in CM than in DM.

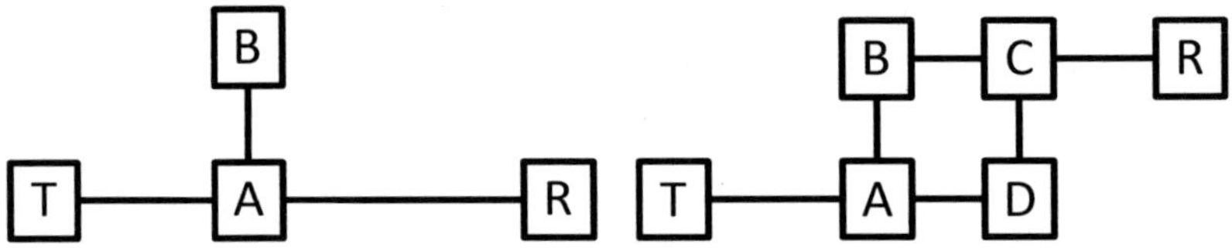

(a) path 1: TAR
path 2: TABAR
path 3: TABABAR
path 4: TABABABAR
...

(b) path 1: TABCR
path 2: TADCR
path 3: TABCDABCR
path 4: TADCBADCR
...

Figure 2.7 Multi-path wave propagation in LV power line networks: T is the transmitter, R is the receiver, A - D are network nodes.

2.2.2 Typical PLC Devices

To use the power lines for communication, various types of PLC devices have been designed. Despite of the differences in implementation, their structures are similar. As introduced in [55], Fig. 2.8 shows a block diagram of a typical PLC modem, which includes a power supply, a coupling circuit, a line driver, an analog front-end (AFE), a physical layer (PHY) controller, a media access control (MAC) unit and an application controller.

The power supply and the coupling circuit share an interface to the grid. Through this grid interface, the power supply module obtains the mains voltage to generate stable DC supplies for the internal circuitries of the modem. The power supply should be clean enough to avoid interferences. It also must show high impedance at the grid side, so that the transmitted signal is not attenuated. On the other hand, the coupling circuit transmits communication signals between the power line network and the modem through the same interface. This coupling circuit must exhibit a band-pass characteristic to let the communication signal pass through, while protecting the modem circuits from damage by the mains voltage.

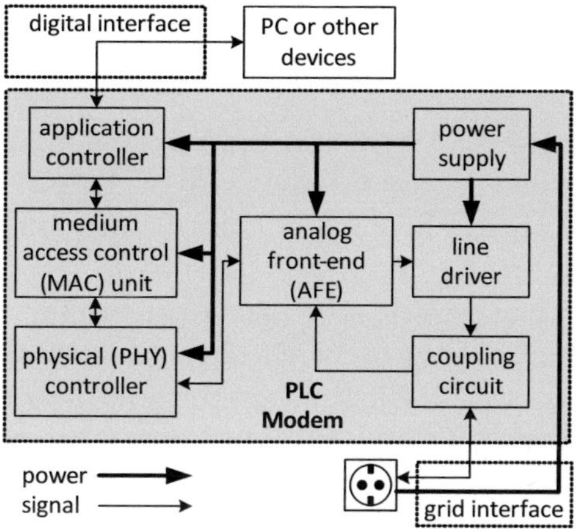

Figure 2.8 A typical PLC modem and its connection to the computer and the grid

The PHY controller, the AFE and the line driver together construct a digital transceiver. Between the PHY controller and the coupling circuit, the AFE plays the role of connecting the digital to the analog worlds. It executes tasks such as analog-to-digital (A/D) and digital-to-analog (D/A) conversions, low-pass filtering, and automatic gain controlling (AGC). Digital processing on the physical layer such as channel coding, error detection/correction, modulation/demodulation and synchronization are accomplished by the PHY controller. The line driver is installed between the coupling circuit and the AFE's transmitting side, in order to amplify the output power so that the strong channel attenuation can be overcome. The MAC unit deals with addressing and channel protocols so that the power network can be shared by multiple nodes for communication without conflict. The application controller accomplishes all tasks above the physical layer and the MAC sublayer. It communicates with a computer or other controlling devices through a digital interface.

2.2.3 Characterizing Power Line Channels: the Bottom-Up and Top-Down Approaches

To develop reliable PLC systems, the first challenge is to characterize and to model power line channels. This task, as the modern PLC technologies are looking forward to higher data rates, has become more important in the recent years. Many efforts have been made since the last decade to build deterministic and statistic models of power line channels [4, 13, 58, 61, 88, 99]. Depending on the approach of channel characterization, these works can be generally divided into two categories: bottom-up and top-down (also know as physical and parametric, respectively [9]).

In the bottom-up approach, structural models of power line channels are built. The cables connecting power consumers and outlets are modeled as two-port networks (2PNs). Each power consumer is modeled as an in parallel connected noise source, which has a complex-valued inner impedance. A typical structural model of this kind is shown in Fig. 2.9. Every 2PN is described by a scattering matrix (S-matrix), which can be obtained in two possible ways. First, if a 2PN can be isolated from the rest of the grid, its S-parameters can be directly measured. Alternatively, but more common, if full physical knowledge about a 2PN (material and geometry of cables, etc.) is available, one can compute its electro-magnetic (EM) properties to obtain its S-parameters. Having all the S-matrices and the knowledge of all loads, both the channel transfer function (CTF) and the overall noise at the receiver can be computed using the TL theory. By changing the network topology, the S-matrices and the loads randomly but reasonably, statistical models can be generated.

The top-down approach, on the other hand, takes a completely behavioral view on power line channels. Ignoring details of wiring and loads, it characterizes each power line channel as a black-box system with a particular access impedance and an additive noise source at the output, as illustrated in Fig. 2.10. The output impedance of the channel is ignored here as usual, due to the fact that most receivers are designed to have very high input impedance. Not only the path loss, but also the access impedance and the additive noise must be obtained via real-world measurements. Usually, parametric models such as proposed in [98, 99] are summarized, based on measurements to simplify the characterization. With a large measurement database, one can obtain the statistics of the model parameters, and thus also a statistical channel model.

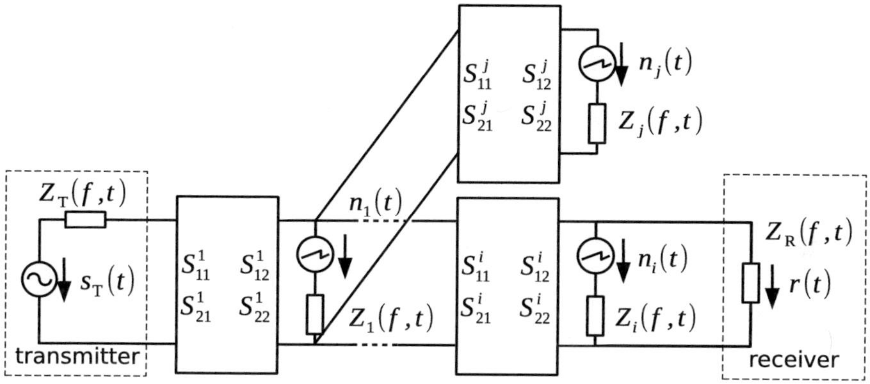

Figure 2.9 Structural model of a power line channel: s_T and r are the transmitted and received signal, respectively. Z_T and Z_R are the impedance of the transmitter and receiver, respectively. Every noise source n_i with the corresponding impedance Z_i models a load at a power outlet.

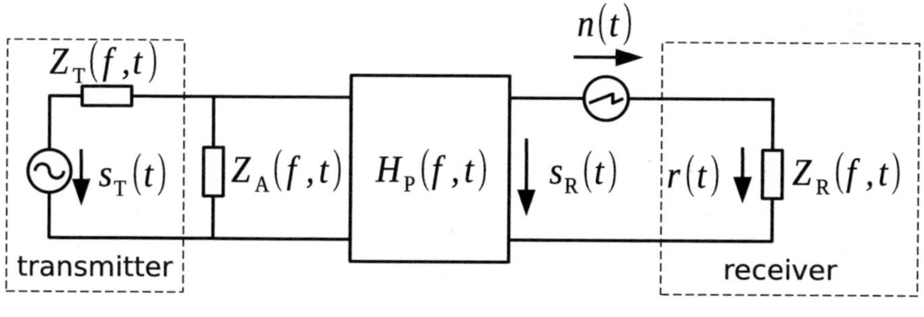

Figure 2.10 Behavioral model of a power line channel: Z_A is the access impedance, H_P is the path loss and n is the additive noise at output. Z_T and Z_R are the impedance of the transmitter and receiver, respectively. s_t and r are the transmitted and received signal, respectively. s_R is the channel response of the input signal.

For existing power line networks, the top-down method is usually preferred. Because full knowledge about the wiring and the loads in such a network is impractical to acquire, so that the structural model cannot be built. On the contrary, the bottom-up approach is superior for hypothetical networks, which are usually analyzed for demonstrating research results. Because it avoids the expensive network implementation and extensive field measurements, which are essential in the top-down approach.

2.2.4 Converting Structural Models into Behavioral Models

As the bottom-up approach is synthetic and the top-down approach is analytic, there is no one-to-one mapping between the structural and behavioral models. Given an arbitrary behavioral model, innumerable different structural models can be found to match it. On the other hand, however, every structural model can be analytically converted into an unique behavioral model. Indeed, behavioral models have also a similar form as structural ones, only predigested into a single 2PN.

Take the simple network in Fig. 2.11(a) for example, where only two segments of cable and one power consumer are connected between the transmitter and the receiver. First, both S-matrices can be converted into their corresponding ABCD matrices. According to the superposition theorem [19], the received voltage signal r equals the sum of the received voltages in Figs. 2.11(b) and 2.11(c). Next, focusing on Fig. 2.11(b), the load impedance Z_1 alone also constructs a simple 2PN, as shown in Fig. 2.11(d). The three cascaded ABCD matrices can be easily merged into one, as shown in Fig. 2.11(e), from which the network's access impedance Z_{A} and the CTF H can be computed, as shown in Fig. 2.11(f). After this, back to Fig. 2.11(b), the output impedance Z_{O}^1 of the 2PN on the left side can be computed to obtain the equivalent network in Fig. 2.11(g). According to Thévenin's theorem [19], this network can be further simplified into the form shown in Fig. 2.11(h). Combining Figs. 2.11(f) and 2.11(h), a behavioral model can be finally obtained in the form shown in 2.10.

Theoretically, given the full information, this conversion also works for any complex network. In practice, due to the high computational complexity of analytical conversions, numerical simulations are usually carried out instead.

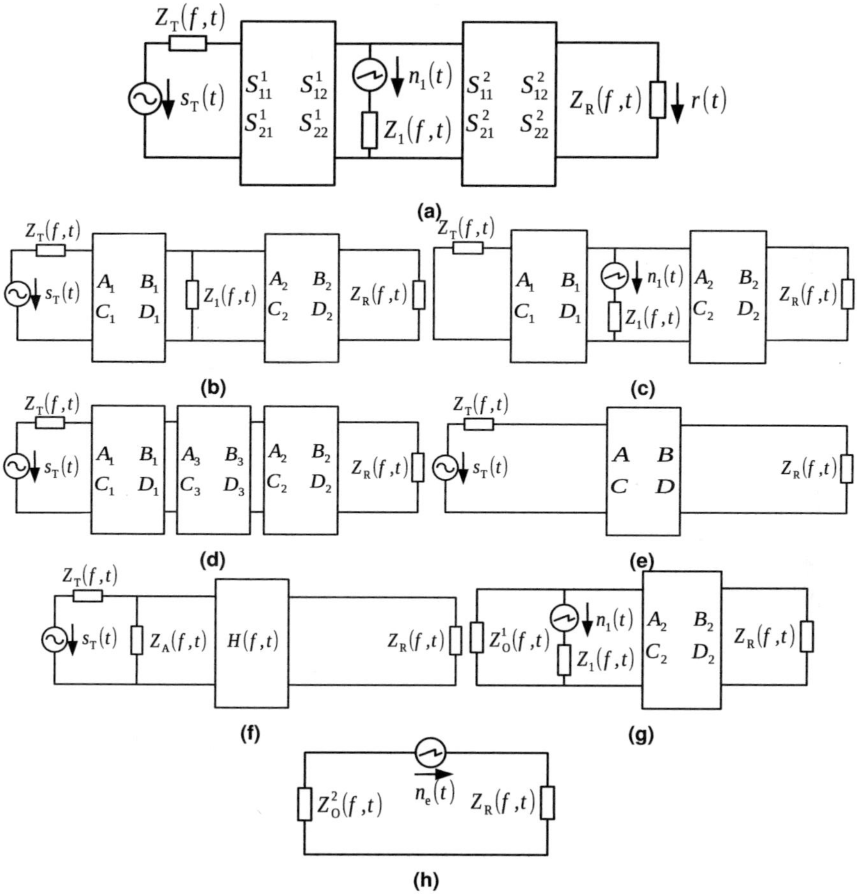

Figure 2.11 Converting the structural model of a simple network into the behavioral model

2.3 Cyclostationary Phenomena within Power Line Channels

Different from the ideal systems in classical system theory, which are linear time-invariant (LTI), and the typical radio propagation channels, which are usually modeled as Rayleigh or Rice fading, the power line channels are not only frequency selective, but also remarkably time-varying. More accurately, they are often referred to as "*cyclic short-term and abruptly long-term time varying*" [9]. The abrupt variation in long terms is mainly caused by random changes of network topology, e.g. turning on/off power consumers or the main breaker of a household. Its impact on the communication can be effectively mitigated through periodic channel estimation and adaptive filtering. The cyclic variation, in contrast, usually happens in terms even shorter than the duration of system adaptation, and therefore can dramatically reduce the quality of communication. This has become the major obstacle for reliable high-speed PLC. In this section, the cyclic time-varying phenomena within power line channels will be introduced, with respect to both CTFs and interferences.

2.3.1 Explanation of Cyclostationary Processes

The cyclic time-varying phenomena within power line channels are usually modeled as cyclostationary processes. To explain this concept, first the definition of stationarity is reviewed as background, then it is generalized into wide-sense local stationarity. At last, the cyclostationarity is defined, and its relations to wide-sense stationarity and wide-sense local stationarity are derived.

2.3.1.1 Stationary and Wide-Sense Stationary Processes

An m-dimensional stochastic process

$$\mathbf{Y}(t) = [Y_1(t), \dots, Y_m(t)]^{\mathrm{T}} \tag{2.1}$$

is called (*strictly* or *strongly*) *stationary*, if its cumulative distribution function (CDF) and probability density function (PDF) are time-invariant (the two expressions are equivalent to each other) [75]:

$$\begin{aligned} &F_{\mathbf{Y}}(y_1,\dots,y_m,t_1,\dots,t_m) \\ =&F_{\mathbf{Y}}(y_1,\dots,x_m,t_1+\tau,\dots,t_m+\tau), \quad \forall t_1,\dots,t_m,\tau \in \mathbb{R}, \end{aligned} \tag{2.2}$$

$$\begin{aligned} &f_{\mathbf{Y}}(y_1,\dots,y_m,t_1,\dots,t_m) \\ =&f_{\mathbf{Y}}(y_1,\dots,y_m,t_1+\tau,\dots,t_m+\tau), \quad \forall t_1,\dots,t_m,\tau \in \mathbb{R}. \end{aligned} \tag{2.3}$$

Or simplified into the one-dimensional version when $m = 1$:

$$F_Y(y,t) = F_Y(y,t+\tau) = F_Y(y), \quad \forall t,\tau \in \mathbb{R}, \tag{2.4}$$

$$f_Y(y,t) = f_Y(y,t+\tau) = f_Y(y), \quad \forall t,\tau \in \mathbb{R}. \tag{2.5}$$

Commonly in signal processing, a weaker version of this concept is applied, known as the *weak* or *wide-sense stationarity*. A signal $y(t)$ is called N^{th}-order wide-sense stationary, if its expectation value and covariance are time-invariant [7, 49]:

$$\begin{aligned} &\int_{-\infty}^{+\infty} y(t) f_y(y(t))\mathrm{d}y \\ =&\mathrm{E}\{y(t)\} = \mathrm{E}\{y(t+\tau)\} = \mathrm{E}\{y\} \quad \forall t,\tau \in \mathbb{R}; \end{aligned} \tag{2.6}$$

$$\begin{aligned} &\mathrm{E}\{(y(t)-\mathrm{E}\{y\})(y(t+\tau)-\mathrm{E}\{y\})\} \\ =&\mathrm{C}_{yy}(t,t+\tau) = \mathrm{C}_{yy}(0,\tau) = \mathrm{C}_{yy}(\tau), \quad \forall t,\tau \in \mathbb{R}. \end{aligned} \tag{2.7}$$

This also implies a time-invariant autocorrelation function (ACF):

$$\begin{aligned} &\mathrm{R}_{yy}(t,\tau) = \mathrm{E}\{y(t)y(t+\tau)\} \\ =&\mathrm{E}\{(y(t)-\mathrm{E}\{y\})(y(t+\tau)-\mathrm{E}\{y\})\} + \mathrm{E}^2\{y\} \\ =&\mathrm{C}_{yy}(\tau) + \mathrm{E}^2\{y\} \\ =&\mathrm{R}_{yy}(\tau) \end{aligned} \tag{2.8}$$

The wide-sense stationarity is always of interest within signal processing, as it is the prerequisite of applying many other important concepts, such as ergodicity. However, many processes in the real world, if not most, do not always satisfy Eqs. (2.6) and (2.7) in the entire time domain. Therefore, a more generalized version of this concept was introduced in [83], known as the *local stationarity in wide-sense.*

2.3.1.2 Wide-Sense Locally Stationary Processes

A stochastic signal $y(t)$ is locally stationary in wide-sense, if

$$\frac{\mathrm{C}_{yy}(t_1,t_2)}{\Gamma_{y,1}(\frac{t_1+t_2}{2})} = \frac{\mathrm{C}_{yy}(t_1+\tau,t_2+\tau)}{\Gamma_{y,1}(\frac{t_1+t_2}{2}+\tau)} = \Gamma_{y,2}(t_2-t_1), \quad \forall t_1,t_2,\tau \in \mathbb{R}, \tag{2.9}$$

where $\Gamma_{y,1}$ is *any non-negative* function of t.

Consider a wide-sense locally stationary signal $y(t)$, of which the moments up to order 2 are time-invariant:

$$\int_{-\infty}^{+\infty} y^n f(y,t)\mathrm{d}y = \mathrm{E}\{y^n(t)\} = M_n, \qquad n = 1,2, \tag{2.10}$$

where M_1 and M_2 are constants independent of t. Obviously, (2.6) is fulfilled ($n = 1$). Letting $\Gamma_{y,1}(t) = \mathrm{E}\{y^2(t)\} = M_2$, (2.7) can also be derived from (2.9). Therefore, it can be asserted that the wide-sense stationarity is a special case of wide-sense local stationarity. Furthermore, even in the context of a time-variant $\Gamma_{y,1}$, $y(t)$ can still be approximated as a wide-sense stationary signal in a short time frame $[t_0-\tau, t_0+\tau]$, if

$$\frac{\mathrm{d}\Gamma_{y,1}(t)}{\mathrm{d}t} < T, \qquad t \in [t_0-\tau, t_0+\tau], \tag{2.11}$$

where T is a threshold.

2.3.1.3 Cyclostationary Processes

Motivated by the interest in the periodic properties of stochastic processes, the concept of cyclostationarity has been intensively studied, since it was proposed before decades. According to [34], a deep review of existing research work on cyclostationary processes, cyclostationarity can be defined in a similar way as stationarity in both strict-sense and wide-sense.

A stochastic process $Y(t)$ is said to be *cyclostationary in strict-sense*, if its CDF (and thus also PDF) is periodic in t with some cyclic period T_0:

$$F_Y(y,t) = F_Y(y,t+NT_0), \quad \forall t \in \mathbb{R}, N \in \mathbb{Z}, \tag{2.12}$$

$$f_Y(y,t) = f_Y(y,t+NT_0), \quad \forall t \in \mathbb{R}, N \in \mathbb{Z}. \tag{2.13}$$

Or as more frequently used in signal processing, a signal $y(t)$ is *second-order cyclostationary in wide-sense*, if its expectation value and ACF are periodic in t with some cyclic period T_0:

$$\mathrm{E}\{y(t)\} = \mathrm{E}\{y(t+NT_0)\}, \quad \forall t \in \mathbb{R}, N \in \mathbb{Z}, \tag{2.14}$$

$$\mathrm{R}_{yy}(t,\tau) = \mathrm{R}_{yy}(t+NT_0,\tau), \quad \forall t,\tau,T_0 \in \mathbb{R}, N \in \mathbb{Z}. \tag{2.15}$$

We can notice that the restriction (2.15) is equivalent to a periodic covariance:

$$\begin{aligned} &\mathrm{C}_{yy}(t,t+\tau) = \mathrm{R}_{yy}(t,\tau) - \mathrm{E}^2\{y\} \\ =&\mathrm{R}_{yy}(t+NT_0,\tau) - \mathrm{E}^2\{y+NT_0\} \\ =&\mathrm{C}_{yy}(t+NT_0,t+\tau), \quad \forall t,\tau,T_0 \in \mathbb{R}, N \in \mathbb{Z}. \end{aligned} \tag{2.16}$$

For convenience, in this thesis, the word *cyclostationary* always means second-order cyclostationary (CS2) in wide-sense, unless explicitly indicated otherwise.

Now consider an arbitrary wide-sense locally stationary signal $y(t)$, of which the moments up to order 2 are periodic:

$$\mathrm{E}\{y^n(t)\} = \mathrm{E}\{y^n(t+NT_0)\}, \qquad \forall t \in \mathbb{R}, N \in \mathbb{Z}, n = 1,2. \tag{2.17}$$

Obviously (2.14) is fulfilled (n=1). Recalling (2.9) and letting

$$\Gamma_{y,1}(t) = \mathrm{E}\{y^2(t)\}, \tag{2.18}$$

(2.15) is also satisfied. Hence, $y(t)$ is cyclostationary.

On the other hand, noting that any constant is periodic in t with arbitrary period, Eqs. (2.6) and (2.7) are only special cases of Eqs. (2.14) and (2.16), respectively.

So far, it can be asserted that:

1. *All wide-sense stationary processes are cyclostationary.*
2. *All cyclostationary processes are wide-sense local stationary.*

Furthermore, cyclostationary signals can be divided into 3 categories:

- class A: signals with non-constant periodic expectation values (periodic signals);
- class B: signals with constant expectation values and non-constant periodic covariance/ACF;
- class C: wide-sense stationary signals.

2.3.2 Cyclostationary Phenomena with Respect to Channel Transfer Functions

As introduced in [26], even when no abrupt change in the topology of a power line network or the attached load happens, the channel transfer function always exhibits a short-term variation, typically with 1/2 of the mains period (i.e. 10 ms in Europe, China, etc.). This is attributed to the fact that the EM characteristics of some electrical devices depend on the instantaneous amplitude of the mains voltage.

Recalling the behavioral model for power line channels in Fig. 2.10, as pointed out in [55], the received voltage signal can be represented in the frequency domain as

$$\begin{aligned} R(f,t) &= S_{\mathrm{R}}(f,t) + N(f,t) \\ &= S_{\mathrm{T}}(f,t) \cdot \frac{Z_{\mathrm{A}}(f,t)}{Z_{\mathrm{A}}(f,t) + Z_{\mathrm{T}}(f,t)} \cdot H_{\mathrm{P}}(f,t) + N(f,t) \\ &= S_{\mathrm{T}}(f,t) \cdot H_{\mathrm{A}}(f,t) \cdot H_{\mathrm{P}}(f,t) + N(f,t). \end{aligned} \tag{2.19}$$

Hence, the overall channel transfer function is

$$H(f,t) = H_{\mathrm{A}}(f,t) \cdot H_{\mathrm{P}}(f,t), \tag{2.20}$$

where $H_{\mathrm{A}}(f,t)$ describes the coupling loss and $H_{\mathrm{P}}(f,t)$ represents the path loss.

There are two solid reasons to assign such an importance to the coupling loss in PLC, that it is isolated from the CTF and separately discussed. First, the access impedance Z_{A} in real power grids is always very low, especially at low frequencies, so that a serious impedance mismatching is present with most common transceivers [91]. Secondly, Z_{A} is time-varying, and may even exhibit a significant periodicity [82].

It has been intensively attempted to mitigate the coupling loss in PLC. To deal with the low Z_{A} in different loading scenarios, a systematic solution to design impedance-adapting couplers was proposed in [91]. To compensate for the periodic time variation in Z_{A}, adaptive algorithms and special coupling circuitries were designed, based on micro-controllers [12, 51, 86]. For communication systems with these features, the influence of H_{A} can be almost neglected and the overall CTF H mainly depends on H_{P}, which shows only slight periodic fluctuation in the frequency band up to 500 kHz [58] but significant periodicity at higher frequencies up to 20 MHz [14]. Otherwise, especially when other appliances are connected to the network near the transmitter, H_{A} can exhibit a strong periodicity [82] and dominate the behavior of H. Generally, H should be modeled to be cyclostationary.

2.3.3 Cyclostationary Phenomena with Respect to Interference

Compared to the periodic fluctuation of CTFs, the cyclostationary phenomena within interferences in PLC are much more significant and have been studied in more detail. It is generally agreed that the interferences in PLC can be classified into four categories[8]:

- narrowband noise (NBN),
- periodic impulsive noise (PIN),
- aperiodic impulsive noise (APIN),
- colored background noise (CBGN).

NBN appears as strong interference over long durations and in narrow frequency bands. There are two kinds of NBN sources. First, electrical appliances, such as switched-mode power supplies, fluorescent lamps and monitors can generate NBNs below 500 kHz. Second, radio signals can be coupled from the air into the grid at power cables, which are long enough to work as antennas. The cyclic behavior of NBN has been reported in many contributions. Influenced by the mains voltage, the first kind of sources, the coupling and the propagation of NBNs over power lines can vary periodically. Thus, NBNs are often amplitude modulated by periodic envelopes, synchronous to the mains [16, 55, 69]. Additionally, a special subclass of NBN called swept frequency noise (SFN) was reported in [55] to be generated by power factor correction (PFC) circuits, and to have periodic central frequencies and bandwidths.

PIN shows the most significant cyclic characteristics among all types of noise. It consists of periodic pulses with short duration (tens of microseconds), but high amplitude (tens of mV), originating from the non-linear circuities connected to the network, whose electrical properties change periodically (e.g. rectifier diodes that are turned on and off by the mains voltage). Depending on the cyclic period, PIN is usually further divided into two classes: synchronous to the mains voltage (PINS, with cyclic period equal to the mains period or two times of it), and asynchronous to the mains voltage (PINAS, with typical cyclic periodicity between 50 and 200 kHz).

In a previous study the cyclostationary characteristics of NBN (excluding SFN) and PIN in LV power line networks have been investigated, and it is concluded that they are cyclostationary according to the classes A and B, respectively [101].

APIN, also known as asynchronous impulsive noise (AIN), corresponds to unpredictable pulses with no regular occurrence. The primary source of APIN is transients during connection and disconnection of electrical devices. Compared to PIN, it occurs more sporadically, exhibits longer pulse durations (up to a few milliseconds) and higher instant power (up to 50 dB above the background level). Most importantly, APIN has no periodicity and arrives at arbitrary instants. Out of the four types of noise, it is the only one which is not cyclostationary.

The remaining parts of PLC interferences are included in the term CBGN, which results from multiple unknown noise sources. Its power is relatively low, compared with the other types of noise, and generally decreases over frequency. Although neglected sometimes in literature [26], a periodic time variance in the power density function of CBGN has been reported in different contributions [16, 55].

In summary, cyclostationary phenomena exist widely in different PLC interference scenarios.

2.4 Narrowband PLC and Broadband PLC

Through the past decades of continuous research and development in the PLC field, some mature technologies and standards have been established for different frequency bands. In this section, first the frequency bands currently occupied by PLC applications are introduced, then the existing PLC technologies and standards are reviewed. Subsequent is an introduction into a recent attempt toward a novel PLC technology which works adaptively in narrowband or broadband applications. Some analysis of the benefits it brings and the challenges is provided.

2.4.1 Used Frequency Bands for PLC

According to [30], existing PLC technologies have exploited frequency bands up to 250 MHz, covering the super low frequency (SLF), ultra low frequency (ULF), very low frequency (VLF), low frequency (LF), medium frequency (MF), high frequency (HF) and ultra high frequency (UHF) bands defined by the *International Telecommunications Union* (ITU). These PLC technologies can be divided into three classes: ultra narrow band (UNB, below 3 kHz), narrowband (NB, 3 to 500 kHz) and broadband (BB, above 1.8 MHz). Due to the extremely limited bandwidth, UNB-PLC works only with a very low data rate (lower than 1 kbps) and only unidirectionally. Although still used in some applications with low requirement of speed, such as AMR, it is no more interesting in recent research. Current interests in PLC focus mainly on NB and BB.

Compared to the mature NB-PLC, BB-PLC still remains as a blueprint. Some standards have been proposed for experimental systems, but yet no sub-band is legally available for industrial or amateur usage. On the contrary, depending on the country/region, the legal usage of NB-PLC is explicitly authorized in particular frequency bands, as listed below:

- The European *Comité Européen de Normalisation Électrotechnique* (CENELEC) bands: 3 to 148.5 kHz, including the A (3 to 95 kHz), B (90 to 125 kHz), C (125 to 140 kHz) and D (140 to 148.5 kHz) bands
- The *United States Federal Communications Commission* (US FCC) band: 10 to 490 kHz
- The Japanese *Association of Radio Industries and Businesses* (ARIB) band: 10 to 450 kHz
- The Chinese band: 3 to 500 kHz.

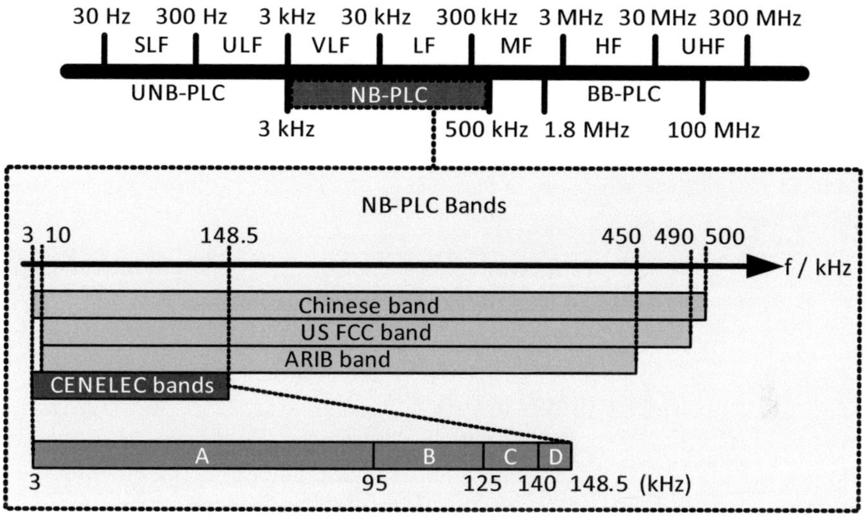

Figure 2.12 Frequency bands exploited by PLC

2.4.2 Existing PLC Standards and Specifications

Standards are established to fix not only the frequency bands, but almost every aspect of PLC on the physical layer and the data link layer (DLL). Up to now, the standard development organizations (SDOs) listed in Table 2.1 have backed some international standards, such as listed in Tab. 2.2. Most modern PLC technologies depend on these standards. It is also common to further classify the NB-PLC technologies into two generations: low data rate (LDR) and high data rate (HDR). The first one represents to the technologies based on single-carrier or spread-spectrum modulation and being capable of data rates of a few kbits/s. The latter refers to technologies based on orthogonal frequency division multiplexing and being capable of data rates up to 500 kbits/s.

Besides these standards, some industrial specifications for PLC have also been released, as listed in Table 2.3. A variety of modulation schemes, such as binary phase shift keying (BPSK), amplitude shift keying (ASK), differential code shift keying (DCSK), and frequency shift keying (FSK) has been applied in LDR technologies, while the modern PLC technologies (HDR, BB) widely combine orthogonal frequency-division multiplexing (OFDM) with phase shift keying (PSK) or quadrature amplitude modulation (QAM).

Table 2.1 Several SDOs supporting the existing international PLC standards and their abbreviations

Full Name	Abbreviation
International Electrotechnical Comission	IEC
International Organization for Standardization	ISO
Commissariat à l'énergie atomique et aux ènergies alternatives	CEA
International Telecommunication Union - Telecommunication Standardization Sector	ITU-T
Institute of Electrical and Electronics Engineers	IEEE
Telecommunications Industry Association	TIA

Table 2.2 Existing PLC standards in NB and BB: The physical signaling (PLS) sublayer is a portion of PHY, the MAC sublayer is a portion of DLL.

Class		Standard	Scope
NB	LDR	ISO/IEC 14908-3	PHY
		ISO/IEC 14543-3-5	PHY, DLL
		CEA-600.31	PHY
		IEC 61334-3-1	PHY
		IEC 61334-5-1	PHY, MAC
		IEC 61334-5-2	PLS, MAC
	HDR	ITU-T G.9902	PHY, DLL
		ITU-T G.9903	PHY, DLL
		ITU-T G.9904	PHY, DLL
		IEEE 1901.2	PHY, MAC
BB		TIA-1113	PHY
		ITU-T G.hn	PHY, DLL
		IEEE 1901	PHY, MAC
		ISO/IEC 12139-1	PHY, MAC

Table 2.3 Several industrial PLC specifications in NB and BB

Class		Specification	Frequency	Modulation	Max. Data Rate[a]
NB	LDR	Insteon	131.65 kHz	BPSK	13 kbps
		X10	120 kHz	ASK	100 bps
		HomePlug C&C	FCC band	DCSK	7.5 kbps
		Ariane Controls	50-3000 kHz[b]	FSK	185 kbps
	HDR	PRIME	42-89 kHz	OFDM DBPSK, DQPSK, D8PSK	128.6 kbps
		G3-PLC	35-91 kHz	OFDM DBPSK, DQPSK	48 kbps
BB		HomePlug AV2	1.8-86.13 MHz	OFDM BPSK, QPSK, 16 to 4096 QAM	200 Mbps
		HomePlug GP	2-30 MHz	OFDM QPSK	10 Mbps
		Gigle MediaXtreme	50-300 MHz	OFDM BPSK, QPSK, 64 to 1024 QAM	882 Mbps

[a] Theoretical maximal raw data rate in the best case

[b] A configuration in this frequency range is technically possible, but practically limited by the local legal PLC band

2.4.3 PLC between 0.15-10 MHz

Between the NB-PLC and BB-PLC specifications, listed in Table 2.3, a significant difference of the maximal achievable data rate can be observed. This is caused by the physical constraint, which is pointed out by the well-known Shannon-Hartley theorem

$$C = B \cdot \log_2(1 + \frac{S}{N}), \tag{2.21}$$

where C is the channel capacity, B is the available bandwidth, S is the average received signal power over the bandwidth, and N is the average noise or interference power over the bandwidth. We can see that under a certain SNR, the channel capacity is proportional to the bandwidth. Therefore, thanks to the higher bandwidth, BB-PLC is able to support faster data transmission.

However, BB-PLC also has its congenital drawback. Due to the loss caused by the cables, power line channels are generally low-pass links. Some high-pass characteristics can be locally observed in narrow bands (e.g. in the CENELEC-A band), but the channels are commonly low-pass links in the broadband applications. Thus, with the same transmitting power, the received signal power S is more attenuated at higher frequencies, which leads to a lower C in any certain bandwidth B. Due to the higher operating frequency range, BB-PLC suffers more from the loss than NB-PLC. Furthermore, the multi-path fading also increases exponentially over the cable length [5]. As the length of a transmission path grows, the losses can become more and more critical, and at last exceed the benefit from higher bandwidth. To overcome the strong attenuation and keep the SNR on a reasonable level, repeaters and relays must be deployed in BB-PLC systems communicating over long distances. These hardware components, however, can intensively raise the cost.

So far, it can be seen that the selection of an appropriate PLC frequency band highly depends on the application requirements. BB-PLC is generally preferred for high-speed communications over short distances (such as "smart home" applications), while NB-PLC adapts low-speed communications over long distances (such as smart metering) better. Beyond this, both the requirements on data rate and the attenuation level can be time-varying in some scenarios, making a band selection even more difficult.

Motivated by this challenge, an attempt toward adaptive band selection between 0.15-10 MHz was made [54]. This new PLC system is intended to select its operating frequency band according to the current channel conditions, instead of any predefined specification. If broadband communication is allowed by the channel, this system works can work in a wide frequency range; otherwise, it finds and uses the best sub-band. With this cognitive technology, the power line channel can be exploied with high efficiency and flexibility. Moreover, the cost of building and running the system can also be reduced.

The implementation and optimization of such a system face two main challenges. First, a model of power line channels should be well defined in the new band of interest (0.15-10 MHz), which crosses the traditional narrowband and broadband approaches. Although previous channel modeling work already fully covers NB and BB, due to the difference in methodology of measuring and characterizing, the results can not be simply taken over to cover the new band. Second, the new system must be tested and evaluated first on the physical layer, before moving to higher levels. However, system test in real power line networks is too inflexible, and emulating power line channels with ordinary laboratory equipment is too inaccurate [55]. Power line channel emulators based on FPGAs have been widely used for flexible laboratory PLC testbeds since the last decade [15, 36, 58, 92], but have not yet been specified for this new band.

3 Measuring and Modeling LV Power Line Channels

To support the development of new PLC systems working between 0.15-10 MHz, power line channels must be well characterized in this frequency band. In the behavioral view, which was introduced in Sec. 2.2.3, a power line channel can be characterized by its transfer function and noise scenario.

Researchers have been measuring the power line channel transfer functions since the beginning of this century [99]. Reported field measurements have covered both the narrowband [48, 58] and the broadband approaches [79]. Most of them only focus on the frequency selectivity of CTFs, or only consider the long-term time variation. However, as analyzed in Sec. 2.3.2, the CTF of a power line channel can also exhibit periodical time variance within a short duration. To measure and model the CTFs more accurately, the existing technologies must be extended.

The noise scenario has also been deeply studied in the past decades. A large mount of measurements has been taken in various regions, and mature models have been built for the observed noises [16, 58, 71, 98]. Nevertheless, the methods used to extract noise of different classes from the measurements are still primitive and lack of accuracy. More advanced approaches are required for a better noise modeling.

3.1 Measuring and Modeling Channel Transfer Functions

3.1.1 Distributed CTF Measuring System

As pointed out in Sec. 2.3.2, the channel transfer function $H(f,t)$ of a power line network can be divided into the coupling loss $H_C(f,t)$ and the path loss $H_\mathrm{P}(f,t)$. As long as the coupling loss is device-dependent, it is more practical to measure the path loss instead of the overall CTF. In this chapter, the expression *CTF* always refers to the path loss $H_\mathrm{P}(f,t)$, if not explicitly commented.

To measure system transfer functions, a vector network analyzer (VNA) is widely used for its wide operating frequency range, high precision and ability to measure phase properties. Unfortunately, VNAs can be hardly used to measure power line channels, mainly due to three reasons. First, the high mains voltage can destroy the vulnerable directional couplers at the VNA's input/output (I/O) ports, so LV couplers must be used to connect the VNA and the power line network. These couplers are usually influenced by the mains voltage and exhibit time-varying EM properties, which makes an accurate VNA calibration very difficult. Second, the input and output ports of power line channels are usually found at different locations with long distances in between, so that long cables are needed to connect the VNA and the LV couplers. According to the dimension of a supply cell, the cable length can range up to hundreds of meters. These long cables further raise the difficulty of calibration and reduce the accuracy of measurements. At last, VNAs commonly have I/O ports fixedly matched to 50 Ohms and only measure S-parameters. Power line channels are, however, never matched, so that the measurement results always include additional coupling losses, and much extra computation is needed to obtain the path loss.

Therefore, distributed CTF measuring systems are usually used instead for power line channel measurements, as illustrated in Fig. 3.1. On the input side is a channel sounder, which is a transmitter that generates a carefully designed signal s_T and sends it into the channel with appropriate power. On the output side there is a measuring receiver, where a voltage signal r_{Mea} is obtained. To exclude the influence of the coupling loss and the behavior of the transmitter, a second receiver is connected to the input, in parallel with the channel sounder, so that the voltage signal r_{Ref} at the channel input is also recorded. According to the behavioral model (2.19) and considering the measurement errors, in the frequency domain there are

$$\begin{aligned}&R_{\text{Mea}}(f,t)\\=&S_{\text{T}}(f,t)\cdot H_A(f,t)\cdot H_{\text{P}}(f,t)+N(f,t)+E_{\text{Mea}}(f,t)\\\triangleq&S(f,t)\cdot H_{\text{P}}(f,t)+N(f,t)+E_{\text{Mea}}(f,t),\end{aligned}\tag{3.1}$$

$$\begin{aligned}&R_{\text{Ref}}(f,t)\\=&S_{\text{T}}(f,t)\cdot H_A(f,t)+N_{\text{Ref}}(f,t)+E_{\text{Ref}}(f,t)\\=&S(f,t)+N_{\text{Ref}}(f,t)+E_{\text{Ref}}(f,t).\end{aligned}\tag{3.2}$$

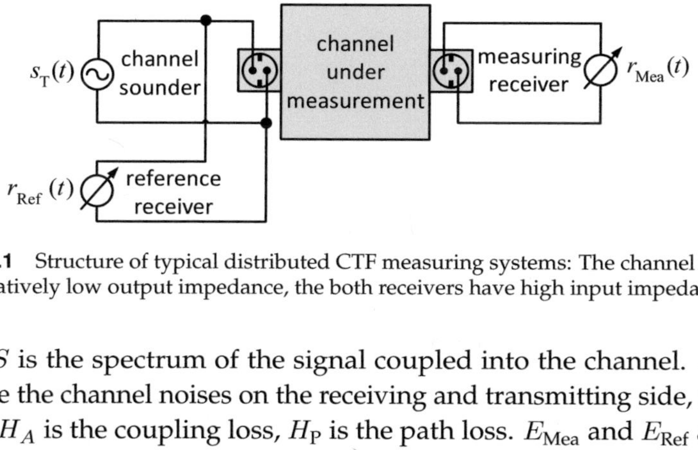

Figure 3.1 Structure of typical distributed CTF measuring systems: The channel sounder has a relatively low output impedance, the both receivers have high input impedances.

Here, S is the spectrum of the signal coupled into the channel. N and N_{Ref} are the channel noises on the receiving and transmitting side, respectively. H_A is the coupling loss, H_{P} is the path loss. E_{Mea} and E_{Ref} are the measurement errors produced by the measuring and reference receivers, respectively, including thermal noises, quantization errors, etc.

In some literature, the reference receiver is absent and the a-priori known s_{T} is used instead of the measured r_{Ref} in the estimation. In this case, the impacts of the coupling loss and the transmitter output impedance are canceled through a calibration [97]. However, considering the time variation of the coupling loss in real networks, an accurate calibration is hard to achieve.

Depending on the selection of the sounding signal s_{T}, different estimators can be designed to estimate the time-varying CTF $H_{\mathrm{P}}(f,t)$.

3.1.2 Two Sounding Signals: Pseudo Noise Sequence and Dual Sweeping Tones

Two types of sounding signals are commonly used for time-invariant channel measurement: the pseudo noise (PN) sequence and dual sweeping tones. A PN sequence is a binary sequence with impulse-like ACF, hence its power is widely distributed in spectrum [74]. By sending a PN sequence through the channel, it is possible to measure the impulse response of the channel and thus to estimate the CTF once entirely. The dual sweeping tones method in contrast is a pair of single-frequency signals that change their frequency over time. By sending a pair of tones through the channel, it is possible to measure the CTF at the frequencies of the tones. As the tones sweep over an extended frequency range, the CTF can be sampled in the entire band of interest. In this subsection, both methods will be extended to measure time-varying channels. Due to the limited time, neither hardware implementation nor real measurement is realized in this work, the methods are only verified with numerical simulations.

3.1.2.1 CTF Estimation Based on PN Sequence

A pseudo noise sequence, usually implemented in the form of a maximal length binary sequence (M-sequence), is *a semi-random sequence in the sense that it appears random within the sequence length, fulfilling the needs of randomness, but the entire sequence repeats indefinitely* [63]. M-sequences are commonly generated with linear feedback shift registers (LFSRs). An M-bit-deep LFSR can generate an M-sequence with the length of $2^M - 1$ bits, as shown in Fig. 3.2.

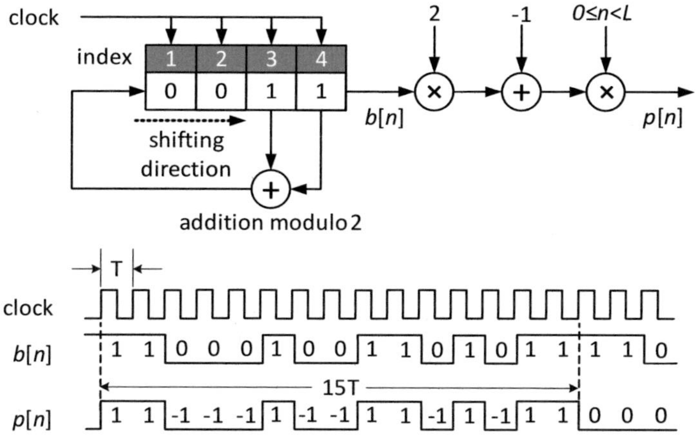

Figure 3.2 Example of an LFSR-based M-sequence generator: $b[n]$ is the M-sequence, which is binary, asymmetrical and periodic over infinite time. $p[n]$ is the symmetrical M-sequence, limited to one sequence length. The depth of the LFSR is $M = 4$, the sequence length is $L = 2^M - 1 = 15$, the clock period is T, the repeating period of $b[n]$ is LT.

According to [70], the deterministic autocorrelation function of the finite-time symmetrical M-sequence $p[n]$ can be computed as

$$C_{pp}[m] = \sum_{n \in \mathbb{Z}} p[n]p[n+m] \qquad \forall m \in \mathbb{Z}. \tag{3.3}$$

As shown in Fig. 3.3, the deterministic autocorrelation function $C_{pp}[m]$ has an impulse-like form. At $m = 0$ it reaches its peak value, which is equal to the sequence length L; with other values of m the magnitude of $C_{pp}[m]$ remains low. With the growth of L, the form of this sequence approaches a Dirac impulse $\delta[m]$, as shown in Fig. 3.4. When L is large enough, it can be approximated that

$$C_{pp}[m] \simeq L\delta[m]. \tag{3.4}$$

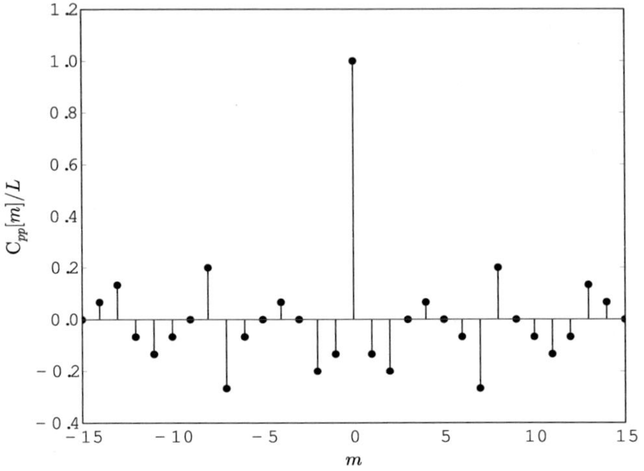

Figure 3.3 The deterministic autocorrelation function of the symmetrical M-sequence $p[n]$ from Fig. 3.2, the sequence reaches its maximum L at $m = 0$.

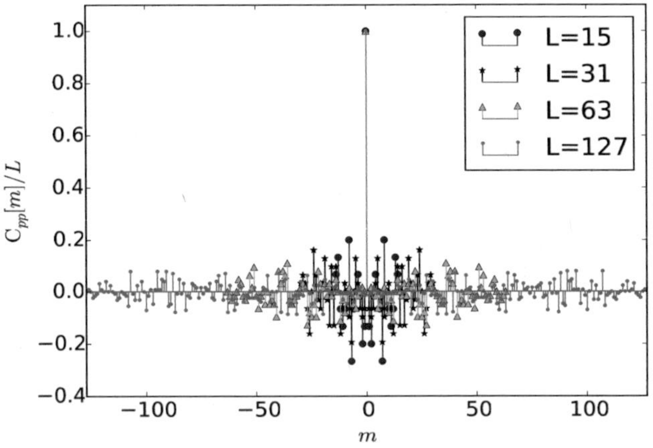

Figure 3.4 The normalized deterministic autocorrelation function $C_{pp}[m]/L$ of some M-sequences with different sequence lengths

Now consider a distributed CTF measurement as shown in 3.1, where the generated sounding signal s_T is shorter than the coherence time of the channel, so that the time variance of the CTF can be neglected during the measurement. Thus Eqs. (3.1) and (3.2) can be simplified in the frequency domain to

$$R_\mathrm{Mea}(f) = S(f) \cdot H_\mathrm{P}(f) + N(f) + E_\mathrm{Mea}(f), \tag{3.5}$$
$$R_\mathrm{Ref}(f) = S(f) + N_\mathrm{Ref}(f) + E_\mathrm{Ref}(f); \tag{3.6}$$

or in the discrete-time domain as

$$r_\mathrm{Mea}[k] = s[k] * h_\mathrm{P}[k] + n[k] + e_\mathrm{Mea}[k], \tag{3.7}$$
$$r_\mathrm{Ref}[k] = s[k] + n_\mathrm{Ref}[k] + e_\mathrm{Ref}[k], \tag{3.8}$$

where $x[k] = x(kT_\mathrm{S})$ with the sampling frequency $1/T_\mathrm{S}$ and $*$ represents the convolution:

$$a[k] * b[k] = \sum_{l=-\infty}^{+\infty} a[k-l]b[l]. \tag{3.9}$$

The coupling loss H_A can be approximated by a constant complex-valued linear scaling

$$H_A = \frac{s(t)}{s_\mathrm{T}(t)} = \frac{Z_A}{Z_A + Z_\mathrm{T}}. \tag{3.10}$$

Thus, the cross-correlation function between r_Ref and r_Mea is

$$\begin{aligned} &\mathrm{R}_{r_\mathrm{Ref} r_\mathrm{Mea}}[l] = \mathrm{E}\{r_\mathrm{Ref}[k] r_\mathrm{Mea}[k+l]\} \\ =&\mathrm{E}\{(s[k] + n_\mathrm{Ref}[k] + e_\mathrm{Ref}[k]) \cdot (s[k+l] * h_\mathrm{P}[k+l] \\ &+ n[k+l] + e_\mathrm{Mea}[k+l])\}. \end{aligned} \tag{3.11}$$

Assuming that a transmitter sends the sounding signal with enough power, the channel noise and the measurement error at the transmitting side can be approximately neglected, so that

$$r_{\text{Ref}}[k] \simeq s[k] \tag{3.12}$$

$$\begin{aligned}
&\mathrm{R}_{r_{\text{Ref}} r_{\text{Mea}}}[l] \\
\simeq &\mathrm{E}\{s[k] \cdot (s[k+l] * h_{\text{P}}[k+l] + n[k+l] + e_{\text{Mea}}[k+l])\} \\
= &\mathrm{E}\{s[k] \cdot (s[k+l] * h_{\text{P}}[k+l])\} + \mathrm{E}\{s[k] n[k+l]\} \\
&+ \mathrm{E}\{s[k] e_{\text{Mea}}[k+l]\}
\end{aligned} \tag{3.13}$$

Assuming that the sounding signal s_{T} is highly uncorrelated with the channel noise n and the measurement error e_{Mea}, due to (3.10) s is also uncorrelated with them. Therefore we get

$$\begin{aligned}
&\mathrm{R}_{r_{\text{Ref}} r_{\text{Mea}}}[l] \simeq \mathrm{E}\{s[k] \cdot [s[k+l] * h_{\text{P}}[k+l]]\} \\
= &\mathrm{E}\{s[k] \sum_{m=-\infty}^{+\infty} s[k+l-m] h_{\text{P}}[m]\} \\
= &\sum_{m=-\infty}^{+\infty} \mathrm{E}\{s[k] s[k+l-m]\} h_{\text{P}}[m] \\
= &\sum_{m=-\infty}^{+\infty} \mathrm{R}_{ss}[l-m] h_{\text{P}}[m]
\end{aligned} \tag{3.14}$$

Assuming that s_{T} is zero-mean, there is

$$\mathrm{R}_{ss}[l] = H_A^2 \mathrm{R}_{s_{\text{T}} s_{\text{T}}}[k] = H_A^2 (\mathrm{C}_{s_{\text{T}} s_{\text{T}}}[l] + \mathrm{E}^2\{s_{\text{T}}\}) = H_A^2 \mathrm{C}_{s_{\text{T}} s_{\text{T}}}[l]. \tag{3.15}$$

Both the assumptions of being uncorrelated with the noise/error and of zero-mean are fulfilled by M-sequences. Hence, an M-sequence with a sufficient length L can be used as a sounding signal s_{T}. Recalling (3.4) there is

$$\mathrm{R}_{ss}[l] \simeq H_{\text{A}}^2 L \delta[l], \tag{3.16}$$

and (3.14) turns into

$$\mathrm{R}_{r_{\mathrm{Ref}}r_{\mathrm{Mea}}}[l] \simeq \sum_{m=-\infty}^{+\infty} H_{\mathrm{A}}^2 L\delta[l-m]h_{\mathrm{P}}[m] = H_{\mathrm{A}}^2 L h_{\mathrm{P}}[l]. \quad (3.17)$$

Similarly it can be derived that the ACF of r_{Ref} is

$$\mathrm{R}_{r_{\mathrm{Ref}}r_{\mathrm{Ref}}}[l] \simeq H_{\mathrm{A}}^2 L\delta[l]. \quad (3.18)$$

Furthermore, assuming that r_{Ref} and r_{Mea} are ergodic, the ACF and the cross-correlation function can be estimated by the inner products averaged over the sequence duration:

$$\hat{\mathrm{R}}_{r_{\mathrm{Ref}}r_{\mathrm{Ref}}}[l] = \frac{1}{L}\sum_{k=0}^{L} r_{\mathrm{Ref}}[k]r_{\mathrm{Ref}}[k+l], \quad (3.19)$$

$$\hat{\mathrm{R}}_{r_{\mathrm{Ref}}r_{\mathrm{Mea}}}[l] = \frac{1}{L}\sum_{k=0}^{L} r_{\mathrm{Ref}}[k]r_{\mathrm{Mea}}[k+l]. \quad (3.20)$$

Thus, the impulse response of the path loss $h_{\mathrm{P}}(t)$ can be estimated as

$$\hat{h}_{\mathrm{P}}[k] = H_{\mathrm{A}}^2 L\hat{\mathrm{R}}_{r_{\mathrm{Ref}}r_{\mathrm{Mea}}}[k] = \hat{\mathrm{R}}_{r_{\mathrm{Ref}}r_{\mathrm{Ref}}}[0]\hat{\mathrm{R}}_{r_{\mathrm{Ref}}r_{\mathrm{Mea}}}[k], \quad (3.21)$$

and the frequency response $H_{\mathrm{P}}(f)$ can be estimated by the discrete Fourier transform (DFT).

A digital implementation of M-sequence-based time-invariant CTF estimation was presented and evaluated by *Zimmermann* in [97], where the method was verified through simulations and experiments. The simulations also showed that the performance can be significantly improved by repeating the measurement several times and taking the average estimation.

Considering the time variation of the CTF, the method must be extended with respect to the coherence time of the channel. As earlier referred to in Sec. 2.3, the power line channels are in short terms cyclic and in long terms abruptly varying, so that three *timescales* can be introduced: the *invariant scale*, the *cyclic scale* and the *random scale* [26, 78], as illustrated in Fig. 3.5. The CTF can be approximately considered as time-invariant within T_{I}, i.e.

$$H(f,t) = H(f,t_0) \qquad \forall t_0 \in \mathbb{R}, t \in [t_0, t_0 + T_{\text{I}}), \tag{3.22}$$

and a periodicity in terms of T_{C} is *statistically* valid within T_{R}, i.e.

$$H(f,t_0) = H(f,t_0 + nT_{\text{C}}) \qquad \forall n \in \{0, 1, \ldots, \lfloor \frac{T_{\text{R}}}{T_{\text{C}}} \rfloor\}. \tag{3.23}$$

According to our investigations, T_{I} is typically in the order of hundreds of microseconds, T_{C} is normally a mains period and T_{R} is many orders of magnitude above T_{C}, sometimes up to several hours [26, 82].

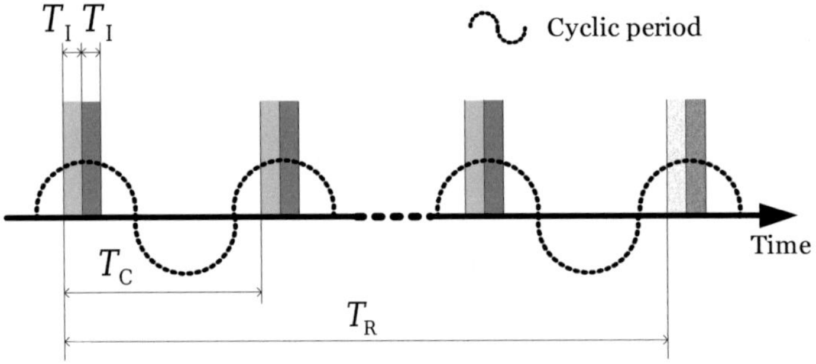

Figure 3.5 Time variations of the power line channel in short and long terms: T_{I} is the length of the invariant scale, T_{C} is the cyclic period and T_{R} is the length of a random scale.

Thus, a series of time sets $\mathbb{S}_k$ can be defined as

$$\begin{aligned} \mathbb{S}_k &= [(k-1)T_{\text{I}}, kT_{\text{I}}) \cup [(k-1)T_{\text{I}} + T_{\text{C}}, kT_{\text{I}} + T_{\text{C}}) \cup \ldots \\ &\quad \cup [(k-1)T_{\text{I}} + NT_{\text{C}}, kT_{\text{I}} + NT_{\text{C}}), \\ N &= \lfloor \frac{T_{\text{R}}}{T_{\text{C}}} \rfloor, k \in \{1, 2, \ldots \lfloor \frac{T_{\text{C}}}{T_{\text{I}}} \rfloor\}, \end{aligned} \tag{3.24}$$

where for each k, the CTF can be considered as time-invariant over $\mathbb{S}_k$.

To sound a channel, the length of the PN sequence can be typically set to the order of dozens of microseconds [97], which is significantly shorter than T_{I}. Thus, with an accurate synchronization, the time-invariant CTF estimator (3.21) can be applied in every $\mathbb{S}_k$, and the time-varying CTF $H_{\mathrm{P}}(f,t)$ can be measured.

3.1.2.2 CTF Estimation Based on Dual Sweeping Tones

Another approach is to directly measure the complex-valued path loss at given frequencies, i.e. to "sample" the path loss in the frequency domain.

First, based on Eqs. (3.1) and (3.2), the magnitude and the phase of the path loss can be estimated as

$$|\hat{H}_{\mathrm{P}}(f,t)| = \frac{|\hat{R}_{\mathrm{Mea}}(f,t)|}{|\hat{R}_{\mathrm{Ref}}(f,t)|}, \tag{3.25}$$

$$\angle\hat{H}_{\mathrm{P}}(f,t) = \angle\hat{R}_{\mathrm{Mea}}(f,t) - \angle\hat{R}_{\mathrm{Ref}}(f,t), \tag{3.26}$$

where $\hat{R}_{\mathrm{Mea}}$ and $\hat{R}_{\mathrm{Ref}}$ are the time-dependent spectral estimations of the signals received by the measuring and reference receivers, respectively.

One straight-forward solution is to sound the channel with the sum of a group of tones

$$s_{\mathrm{T}}(t) = \sum_{k=1}^{N} A_k \cos(2\pi f_k t - \varphi_k), \tag{3.27}$$

so that the SNRs at the receiver $\mathrm{SNR}_R(f,t)$ have peaks at the tone frequencies $f_1, f_2, \dots, f_N$, where $|\hat{H}_{\mathrm{P}}(f,t)|$ can be estimated with high precision. An advantage of this multi-tone method is that the path loss at all frequencies of interest can be obtained with a fixed sounding signal. It has been successfully applied in narrowband channel measurements [85]. However, the estimation accuracy reduces linearly with an increasing number of tones. Even with the same number of tones, due to the stronger attenuation of power line channels at higher frequencies, more driving capacity of the transmitter is needed for measurements in broadband applications.

Furthermore, for an accurate phase estimation, a perfect synchronization between the reference and measuring receivers is essential, which is usually hard to achieve.

To realize accurate CTF measurements up to 10 MHz with high frequency resolution, the dual sweeping tones proposed in [55] can be employed to sound the channel. Instead of consistently sending many tones together, the transmitter sounds the channel several times, i.e. each time it sends only two tones with a fixed frequency difference (the so-called *aperture*) Δf, and the frequencies shift by Δf per time step, i.e.

$$s_{\mathrm{T},i}(t) = A_i \cos(2\pi f_i t - \varphi_i) + A_{i+1}\cos(2\pi f_{i+1} t - \varphi_{i+1}), \tag{3.28}$$

$$f_{i+1} = f_i + \Delta f, \tag{3.29}$$

where i is the index of the tone pair, so that $|\hat{H}_{\mathrm{P}}(f_i,t)|$ and $|\hat{H}_{\mathrm{P}}(f_{i+1},t)|$ can be estimated by (3.25). The group delay can be approximated as

$$\hat{\tau}_{\mathrm{P}}(f_i,t) \simeq \frac{\hat{R}_{\mathrm{Mea}}(f_i,t) - \hat{R}_{\mathrm{Ref}}(f_1,t) - \hat{R}_{\mathrm{Mea}}(f_{i-1},t) + \hat{R}_{\mathrm{Ref}}(f_{i-1},t)}{-360^\circ \cdot \Delta f}. \tag{3.30}$$

Recalling (3.23), for any i, it is equivalent to measure $H_{\mathrm{P}}(f_i,t)$ and to measure $H_{\mathrm{P}}(f_i,t+nT_{\mathrm{C}})$, when $nT_{\mathrm{C}} < T_{\mathrm{R}}$. Thus, the CTF measurement can be divided into K sub-measurements, synchronized with the cyclic period, as illustrated in Fig. 3.6. In the ith sub-measurement, the sounding tone-pair $s_{\mathrm{T},i}$ is transmitted, so that the time-varying magnitude response $|H_{\mathrm{P}}(f,t)|$ and group delay $\tau_{\mathrm{P}}(f,t)$ of the channel at the frequency f_i can be measured over N cyclic periods, according to Eqs. (3.25) and (3.30), respectively. After K sub-measurements, both $|H_{\mathrm{P}}(f,t)|$ and $\tau_{\mathrm{P}}(f,t)$ are sampled over the frequency range $[f_0, f_{K-1}]$ with a resolution of Δf. And the phase response can be estimated as

$$\angle \hat{H}_{\mathrm{P}}(f_i,t) = \varphi_0 - \pi\Delta f \sum_{k=1}^{i} \hat{\tau}_{\mathrm{P}}(f_k,t). \tag{3.31}$$

Note that the overall measuring duration NKT_{C} should not exceed the random scale T_{R}, so that the approximation (3.23) can be taken.

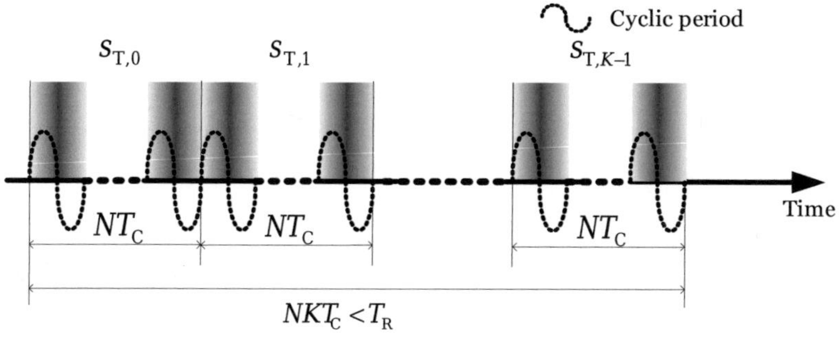

Figure 3.6 Sweeping sounding tones according to the cyclic period

Compared with the multi-tone method, only two tones are transmitted at a time, so that a higher SNR can be achieved with the same hardware, which enables accurate measurement, even with strong channel attenuation and high noise power.

3.1.2.3 Performance Comparison

Both PN sequences and dual sweeping tones have been validated for time-invariant power line channel measurements under proper configurations [55, 97], but no performance comparison between the them has been reported yet. When considering the time variation, configurations of the sounding signal, such as PN sequence length and tones aperture are more strictly constrained. Therefore, it is worth to compare both methods under the same conditions, such as measuring time, transmitting power and frequency resolution. Furthermore, both methods are designed based on the assumption of uncorrelated white noise. However, in real power line channels, this assumption is invalid and the measuring performance therefore sinks. In the following part of this subsection, the two sounding methods are briefly analyzed with simulations and compared with respect to their performance under non-ideal channel conditions. The magnitude response is focused on, because it is more critical than the phase response for communication systems. Additionally, the impact of non-linear hardware behavior is also discussed.

Robustness to Background Noise

We consider a cyclic channel with invariant scale $T_{\mathrm{I}} = 100\,\mu\mathrm{s}$, cyclic period $T_{\mathrm{C}} = 20\,\mathrm{ms}$ and random scale $T_{\mathrm{R}} >> 5\,\mathrm{s}$. In each cyclic period, the channel varies among up to 200 different states. Simulations were executed to measure the CTF with both methods. To simplify the computation, only one state was estimated and analyzed, where the CTF was generated according to sample network in [97]. Zero-mean additive white Gaussian noise (AWGN) was generated as channel noise for simplification. The parameters of the methods with PN sequence and sweeping tones were configured as listed in Tables 3.1 - 3.2, so that the transmitting power and the overall measuring time were uniform in both methods. Testing both methods with different values of noise power, the results shown in Fig. 3.7 were obtained. It can be stated that both methods perform well under the AWGN, even in presence of strong noise and high attenuation. Although the power of a PN sequence is widely distributed over the entire frequency range, which leads to a lower SNR in each measurement than the sweeping tones method, the short period of the sequence allows a large mount of repetitions. Through averaging, the power of AWGN is reduced, so that the SNR is enhanced.

Table 3.1 Basic parameters of the measurements

Parameter	Value	
Synthesis frequency	50 MHz	
Measuring time	8 s	
Transmitting power	$50\,\mathrm{V}^2$	
Frequency resolution	*PN sequence*	24.4 kHz
	Sweeping tones	25 kHz

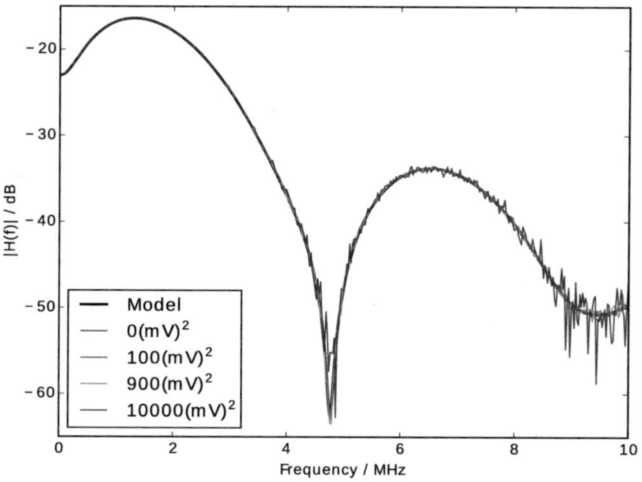

(a) PN sequence method

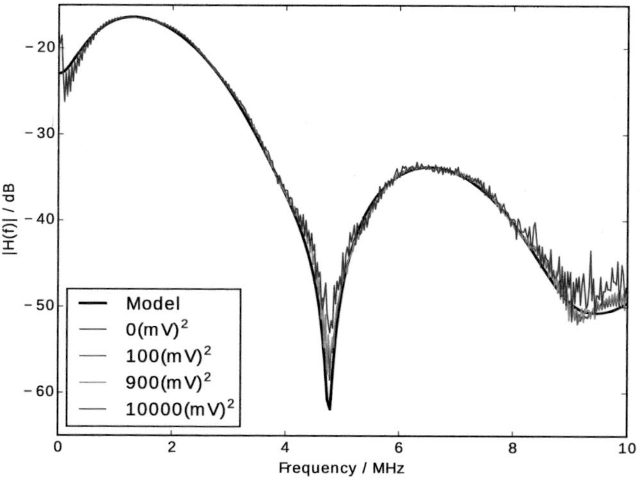

(b) Dual sweeping tones method

Figure 3.7 Simulation results: CTF measurements under AWGN of different power

Table 3.2 Other parameters of the sounding signals

PN Sequence		*Sweeping tones*	
Parameter	**Value**	**Parameter**	**Value**
Sequence length	2047 bits	Aperture	25 kHz
Sequence period	5094 samples	Window length	1024 points
Period length	101.88 µs	Window type	Hamming
		FFT length	1024 points
		FFT overlap rate	63/64

Robustness against Narrowband Noise

In real interference scenarios of power line channels, the boradband background noise is generally relatively weak, while narrowband noises play an important role. They usually exhibit significantly higher amplitudes than the background noise (which will be further discussed in Sec. 3.2) and their presence can remain constant over hours. To investigate the robustness of both methods against NBN, the aforementioned simulations were repeated under a weak AWGN and with two strong NBNs at the measuring receiver. The NBNs were synthesized as amplitude modulated (AM) signals with the parameters listed in Table 3.3. The simulated results are shown in Fig. 3.8. The PN sequence method was slightly more degraded by the NBNs than the dual sweeping tones method, but a median filter of order 7 efficiently suppressed the errors caused by the NBNs for both methods.

Table 3.3 Parameters of the simulation

Parameter	Value
Carrier frequency of the 1st AM signal	900 kHz
Carrier frequency of the 2nd AM signal	4.5 MHz
Bandwidth of the AM signals	10 kHz
AM modulation index	0.2
Power of each AM signal	5000 (mV)^2
Power of the AWGN	100 (mV)^2

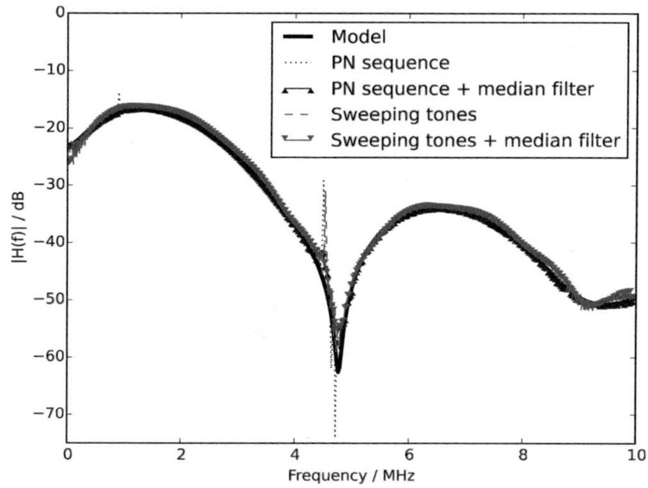

Figure 3.8 Simulation results: CTF measurements in presence of NBN: A median filter of order 7 is applied for smoothing.

Robustness to Non-Linear Hardware

To ensure the accuracy of measurements, a high transmitting power is needed. However, this often overloads the transmitter hardware and leads to non-linear behavior. The sounding signal can be therefore distorted by harmonics. To investigate the impact of unwanted harmonics on the measurements, simulations were executed, where the sounding signals were distorted by second order harmonics of different levels. The noise was synthesized as WGN of 900 $(\mathrm{mV})^2$. The results are shown in Fig. 3.9. Both methods exhibit satisfying robustness to the harmonics.

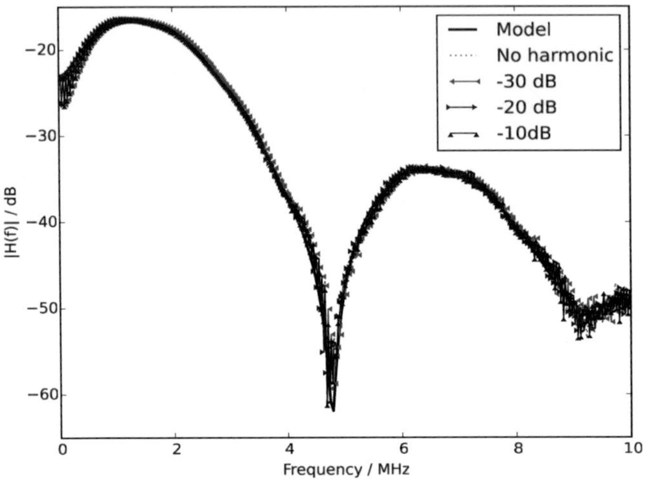

(a) PN sequence method

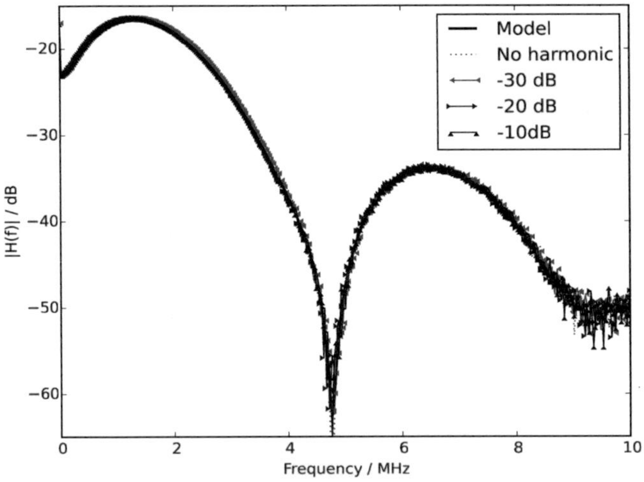

(b) Dual sweeping tones method

Figure 3.9 Simulation results: CTF measurement under presence of 2nd order harmonics

3.1.3 Synchronization

In a distributed power line CTF measuring system, two kinds of synchronizations are required. First, the reference receiver and the measuring receiver must be accurately synchronized, in order to execute the correlated measurements, especially when the phase response is of interest. This can be realized with help of a global positioning system (GPS) receiver on each side, which provides time information accurate to 0.1 μs [55]. Second, as long as the cyclic behavior of the CTF is synchronous to the mains period, the measurement must be synchronized with the mains as well. To solve this problem, a zero-crossing detector (ZCD) can be applied to obtain the phase information of the mains voltage at the receiver, according to its zero-crossing time [103], as shown in Fig. 3.10. It is worth pointing out that the mains phase differs between the transmitting side and the receiving side, depending on the distance and the network topology. Nevertheless, for certain channel, this phase difference remains constant within the random scale, so the mains phase must be measured only at one side.

3.1.4 Setup and Calibration

As discussed above, a complete setup of a distributed CTF measuring system should include a channel sounder, two receivers synchronized with GPS, and a ZCD. More in depth, like any PLC device introduced in Sec. 2.2.2, each receiver must include an AFE and a coupling circuit, which have no ideal all-pass characteristics, as shown in Fig. 3.11. Therefore, instead of $H(f,t)$ the uncalibrated measurement result is

$$H_{\mathrm{Unc}}(f,t) = H_{\mathrm{Ref}}(f) \cdot H(f,t) \cdot H_{\mathrm{Mea}}(f). \tag{3.32}$$

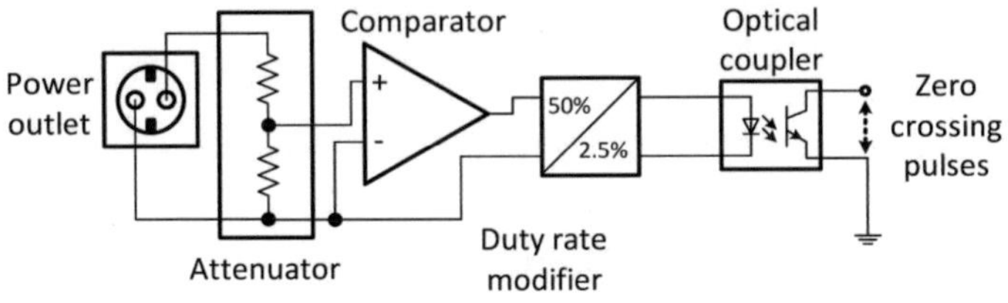

(a) Schematic diagram [103]

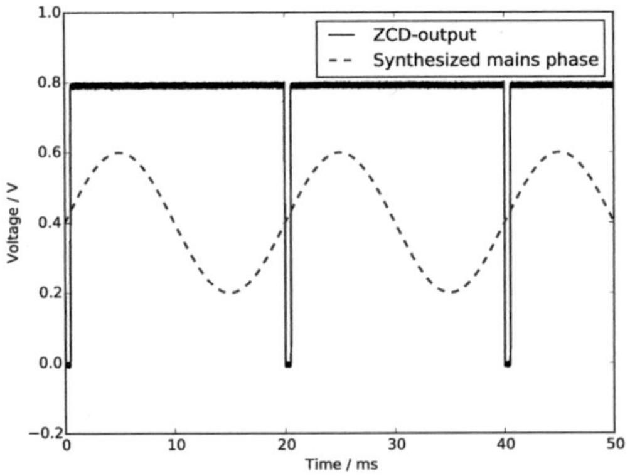

(b) Output with respect to the mains phase, pulse width = 0.5 ms, edge width < 10 µs

Figure 3.10 An off-the-shelf design of a zero-crossing detector

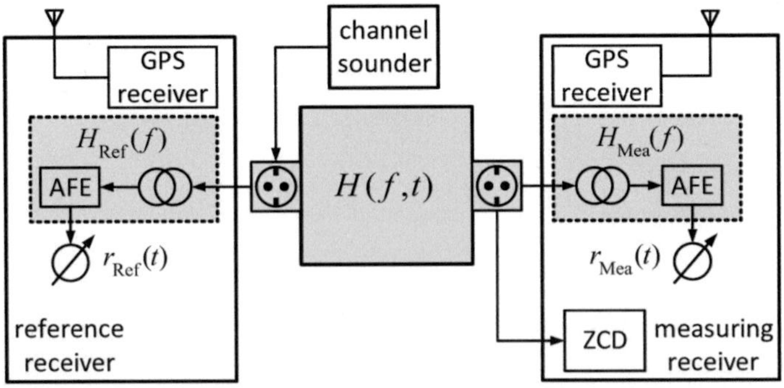

Figure 3.11 Complete setup of a distributed CTF measuring system

To mitigate this system error, a calibrating measurement should be executed with the channel sounder and both receivers directly connected together. Thus, an ideal system $H_{\text{ideal}}(f,t) = 1$ is measured. The calibration result is then

$$H_{\text{Cal}}(f) = H_{\text{Ref}}(f) \cdot H_{\text{Mea}}(f), \tag{3.33}$$

so that the system error can be calibrated through post-processing, i.e.

$$H(f,t) = \frac{H_{\text{Unc}}(f,t)}{H_{\text{Cal}}(f)}. \tag{3.34}$$

3.1.5 Modeling Channel Transfer Functions

3.1.5.1 Linear Time-Invariant Models

It had been common to ignore the time variation and to describe the CTF of power line networks with LTI models, because the short-term fluctuation of the CTF over time influences the communication only slightly, when the data transmission rate is low. Therefore, a long-term averaged channel information was sufficient. Recalling (3.22), power line channels can also be considered as LTI systems when investigated on the invariable scale. Existing LTI models for power line channels can be generally divided into two groups: the multi-path models and the resonant circuit models.

Multi-Path Models

Multi-path models are also usually known as echo models; a typical example is proposed by *Philipps* in [72]:

$$H(f) = \sum_{\nu=1}^{N} |\rho_\nu| \cdot e^{j\varphi_\nu} \cdot e^{-j2\pi f \tau_\nu}, \tag{3.35}$$

where N is the number of paths, $|\rho_\nu|$, φ_ν and τ_ν denote the attenuation, the phase offset and the delay of the νth transmission path, respectively. A more general echo model is proposed by *Zimmermann* in [96]:

$$H(f) = \sum_{i=1}^{N} g_i \cdot e^{-(a_0 + a_1 f^k) d_i} \cdot e^{-j2\pi f \frac{d_i}{v_P}}, \tag{3.36}$$

where N is the amount of paths, a_0 and a_1 are attenuation parameters, k is the exponent of the frequency, usually between 0.2 and 1, v_P is the phase speed, g_i and d_i are the weighting factor and the length of the ith path, respectively. With any given value of $N \geq 1$, the parameters of both models can be estimated from CTF measurements using numerical algorithms. The accuracy rises with growing N.

Resonant Circuit Models

Resonant circuit models attempt to reproduce the measured CTF $H(f)$ with cascaded resonant circuits, i.e.

$$H(f) = \prod_{i=1}^{N} H_i(f), \tag{3.37}$$

where N is the model order and $H_i(f)$ is the system function of the ith reconant circuit, as shown in Fig. 3.12(a).

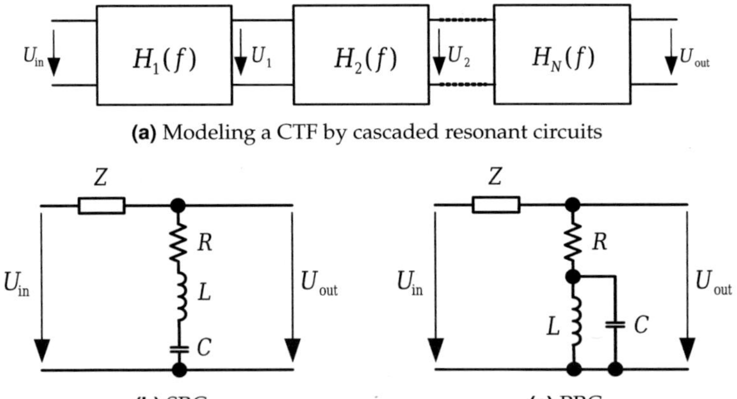

(a) Modeling a CTF by cascaded resonant circuits

(b) SRC

(c) PRC

Figure 3.12 Resonant circuit models

An example series resonant circuit (SRC) model is proposed by *Philipps* in [72], where each $H_i(f)$ is modeled as an SRC, which includes a resistance R, an inductance L and a capacitance C, as illustrated in Fig. 3.12(b). An SRC behaves as a band-stop filter with the resonant frequency $f_{\text{res}} = 1/2\pi\sqrt{LC}$. The access impedance Z must be defined. Given a real CTF measurement with any selected value of $N \geq 1$, the parameters R, L and C can be numerically estimated by means of an evolutionary strategy, and the estimation accuracy rises with growing N [72]. *Liu* improved this model further by adding parallel resonant circuits (PRCs) to the cascade [55]. A PRC has the structure illustrated in Fig. 3.12(c) and behaves as a band-pass filter. The influence of loads on the resonant circuits is also discussed in [55].

3.1.5.2 Linear Periodically Time-Varying Models

When considering the periodical time variation over a cyclic period, a power line channel is generally modeled as linear periodically time-varying (LPTV) [14, 55, 78], i.e. its impulse response varies periodically with time, i.e.

$$h(t,\tau) = h(t,\tau - nT_0), \forall n \in \mathbb{Z}, \tag{3.38}$$

where T_0 is the period. Recalling (3.23), this is fulfilled in power line channels within a random scale.

Volterra Series Model

Liu has proved in [55] that an LPTV power line channel can be expressed in the time domain as a Volterra series in the form

$$\begin{aligned} h(t,\tau) &= h_1(\tau) \\ &+ \sum_{n=2}^{+\infty} \int\int \dots \int h_n[t,\tau_1,\dots,\tau_{n-1}] \prod_{i=1}^{n-1} m(t-\tau_i)\mathrm{d}\tau_1\mathrm{d}\tau_2\dots\mathrm{d}\tau_{n-1}, \end{aligned} \tag{3.39}$$

where $h_n(t_1, t_2, \ldots, t_n)$ is the nth Volterra kernel and $m(t)$ is a periodic function of time. However, not only the estimations of the kernels and $m(t)$, but also synthetic implementation of this model can be computationally complex.

Extended Multi-Path Model

Modeling the channel with a multi-path model and estimating the model parameters in each invariant time interval, the periodical fluctuation of the CTF over time leads to a periodicity for the multi-path model parameters. Thus, an LTI multi-path model can be extended into its LPTV version, then e.g. (3.36) modifies to

$$H(f,t) = \sum_{i=1}^{N} g_i(t) \cdot e^{-[(a_0(t)+a_1(t)f^k(t)]d_i(t)} \cdot e^{-j2\pi f \frac{d_i(t)}{v_P}} . \qquad (3.40)$$

In a given power line network, the attenuation parameters a_0, a_1, the exponent of frequency k, the path lengths d_i and the phase velocity are physically determined by the network topology and the cable type. Thus, they do not vary over time within the random scale. The periodicity of the CTF is caused by the periodical behavior of the reflection and transmission factors of the different paths, which are presented by the factors g_i. Thus, (3.40) can be simplified to

$$H(f,t) = \sum_{i=1}^{N} g_i(t) \cdot e^{-(a_0+a_1 f^k)d_i} \cdot e^{-j2\pi f \frac{d_i}{v_P}} , \qquad (3.41)$$

where $g_1(t), g_2(t) \ldots g_N(t)$ are periodic in terms of the cyclic period T_C. An example of this model is shown in Fig. 3.13, where the model uses the parameters listed in Table 3.4. Extended multi-path models make use of the existing LTI methods, and it is easy to prove that the computational complexity is only linearly related to the time resolution.

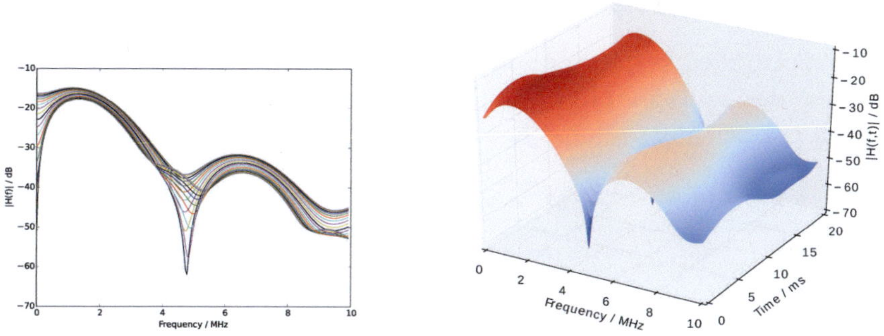

(a) 40 different CTFs in a mains period **(b)** The channel in a time-frequency view

Figure 3.13 An LPTV channel simulated with the extended multi-path model (3.41)

Table 3.4 Model parameters of the simulation in Fig. 3.13.

Parameter	Unit	Value
N	N/A	5
k	N/A	1
v_P	$\mathrm{m \cdot s^{-1}}$	1.53×10^8
a_0	$\mathrm{m^{-1}}$	0
a_1	$\mathrm{s \cdot m^{-1}}$	2.5×10^{-9}
$[d_1, d_2, d_3, d_4, d_5]$	m	$[90.1, 101.8, 113.2, 143, 148]$
g_1	N/A	$0.029 + 0.01 \sin(2\pi t/T_\mathrm{C})$
g_2	N/A	$0.043 + 0.02 \sin(2\pi t/T_\mathrm{C})$
g_3	N/A	$0.103 + 0.1 \sin(2\pi t/T_\mathrm{C})$
g_4	N/A	$-0.058 + 0.02 \sin(2\pi t/T_\mathrm{C})$
g_5	N/A	$-0.046 + 0.02 \sin(2\pi t/T_\mathrm{C})$
T_C	s	20×10^{-3}

3.1.5.3 Linear Switching Time-Varying Model

For a simplified implementation, an LPTV system $H(f,t)$ can be also represented with a set of LTI filters and a switch at the input or output, which periodically switches to one of the LTI filters, i.e.

$$h(t,\tau) = \begin{cases} h_1(t), & \tau \in [\tau_1 + nT_C, \tau_2 + nT_C) \\ h_2(t), & \tau \in [\tau_2 + nT_C, \tau_3 + nT_C) \\ \cdots \\ h_K(t), & \tau \in [\tau_K + nT_C, \tau_1 + (n+1)T_C) \end{cases} \quad \forall n \in \mathbb{Z}, \quad (3.42)$$

where K is the size of the LTI filter bank, τ_i is the time when the input/output switches to the ith LTI filter. Known as linear switching time-varying (LSTV) systems [60], as shown in Fig. 3.14, these structures have been applied to time-variant synthesis of power line channels [15, 78].

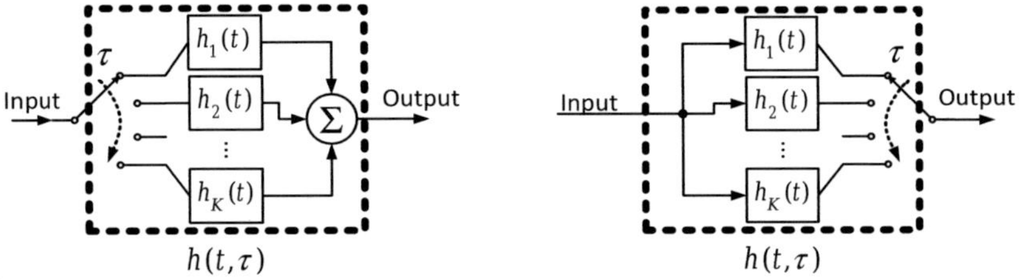

(a) Switch at the input **(b)** Switch at the output

Figure 3.14 Two structures of LSTV systems

Similar to the extended multi-path model, the LSTV model also simplifies the modeling of a time-variant system to the modeling of several LTI systems. But instead of the time functions for model parameters, in the LSTV model the size K of the filter bank and the switching time points $\tau_1, \tau_2 \ldots \tau_K$ are of interest. Considering an LPTV channel estimated under a high time resolution, where N time-invariant CTFs are sampled over a cyclic period and each is estimated with M frequency pins, the CTF can be discretely represented in the matrix

$$\mathbf{H} = \begin{bmatrix}\mathbf{H}_1 & \mathbf{H}_2 & \dots & \mathbf{H}_N\end{bmatrix} = \begin{bmatrix} H_1[1] & H_2[1] & \dots & H_N[1] \\ H_1[2] & H_2[2] & \dots & H_N[2] \\ \vdots & \vdots & \ddots & \vdots \\ H_1[M] & H_2[M] & \dots & H_N[M] \end{bmatrix}, \tag{3.43}$$

where $\mathbf{H}_n$ is the discrete vector representation of the nth invariant CTF $H_n(f)$, and

$$H_n[m] = H(f = m\Delta f, t = n\Delta t), \tag{3.44}$$

where Δf and Δt are the frequency and time resolutions, respectively. The most straight-forward solution is to set $K = N$, so that an LTI filter bank with N filters has to be implemented. However, in real measurements, the channel can vary slowly over time, and thus some invariant CTFs can be very similar to their neighbors. In this case, in order to reduce the implementation cost, a smaller $K < N$ can be taken, and the N invariant CTFs can be clustered into K groups $\{\mathbb{A}_1, \mathbb{A}_2, \dots, \mathbb{A}_i\}$. For each $\mathbb{A}_i$ an LTI filter can be implemented according to the mean CTF

$$\bar{H}_{\mathbb{A}_i}(f) = \frac{1}{N}\sum_{n=1}^{N_i} H_n(f), \tag{3.45}$$

where N_i is the size of $\mathbb{A}_i$. To accomplish the clustering task, two different time-slot segmenting methods are available.

Uniform Segmentation

Assuming that the time fluctuation of the CTF is slow and smooth, a simple solution is to uniformly segment every cyclic period of the CTF into a fixed number K_{fixed} of time slots with the same length, so that these N CTFs are divided into $\hat{K} = K_{fixed}$ groups as follows:

$$\mathbb{A}_i = \{H_{(i-1)A+1}, H_{(i-1)A+2}, \dots, H_{iA}\}, \forall i \in \{1, 2, \dots \hat{K} - 1\} \tag{3.46}$$
$$\mathbb{A}_{\hat{K}} = \{H_{(\hat{K}-1)A+1}, H_{(\hat{K}-1)A+2}, \dots, H_N\} \tag{3.47}$$
$$A = \lfloor N/K_{\text{fixed}} \rfloor. \tag{3.48}$$

The switching time points are also uniformly distributed, i.e.

$$\tau_i = (i-1)A\Delta t, \qquad \forall i \in \{1, 2, \dots, \hat{K}\}. \tag{3.49}$$

This method is simple to implement, but lacks of accuracy and flexibility, when the CTF $H(f,t)$ varies unevenly over time.

Deviation-Based Clustering

To achieve more accuracy and flexibility in the segmentation of time slots, an adaptive clustering approach based on CTF deviation and k-means algorithm is developed.
First, given two time-invariant CTFs $H_1(f)$ and $H_2(f)$ over the frequency range of interest $[f_{\text{begin}}, f_{\text{end}}]$, their magnitude distance and phase distance are defined as

$$M_{H_1H_2} = \frac{1}{f_{\text{end}} - f_{\text{begin}}} \int_{f_{\text{begin}}}^{f_{\text{end}}} \sqrt{(|H_1(f)| - |H_2(f)|)^2} \mathrm{d}f, \tag{3.50}$$
$$P_{H_1H_2} = \frac{1}{f_{\text{end}} - f_{\text{begin}}} \int_{f_{\text{begin}}}^{f_{\text{end}}} \sqrt{|\angle H_1(f) - \angle H_2(f)|^2} \mathrm{d}f. \tag{3.51}$$

Next, given a set of time-invariant CTFs $\mathbb{A} = \{H_1(f), \dots, H_N(f)\}$, the deviation of $\mathbb{A}$ is defined as

$$\sigma_{\mathbb{A}}' = \frac{1}{N} \sum_{n=1}^{N} [wM_{H_n\bar{H}_{\mathbb{A}}} + (1-k)P_{H_n\bar{H}_{\mathbb{A}}}], \tag{3.52}$$

where w is the weighting factor of the magnitude distance.

For the discrete LPTV channel measurement 3.43, an observation matrix can be constructed as follows:

$$\mathbf{F} = \left[\mathbf{F}_1\,\mathbf{F}_2\ldots\mathbf{F}_N\right] = \begin{bmatrix} |H_1[1]| & |H_2[1]| & \ldots & |H_N[1]| \\ |H_1[2]| & |H_2[2]| & \ldots & |H_N[2]| \\ \vdots & \vdots & \ddots & \vdots \\ |H_1[M]| & |H_2[M]| & \ldots & |H_N[M]| \\ \angle H_1[1] & \angle H_2[1] & \ldots & \angle H_N[1] \\ \angle H_1[2] & \angle H_2[2] & \ldots & \angle H_N[2] \\ \vdots & \vdots & \ddots & \vdots \\ \angle H_1[M] & \angle H_2[M] & \ldots & \angle H_N[M] \end{bmatrix}, \tag{3.53}$$

where $\mathbf{F}_n$ is the observation vector of $H_n(f)$.

Applying the k-means algorithm [59] to the observation matrix $\mathbf{F}$, the N invariant CTFs can be clustered into k sets: $\{\mathbb{A}_1, \ldots \mathbb{A}_k\}$. Iteratively increasing the parameter k from 1 on and repeating the clustering, until every cluster has a deviation lower than a given threshold $\sigma_{\max}$, we can get

$$\hat{k} = \arg\min_{k} \sigma_{\mathbb{A}_i} \leq \sigma_{\max}, \qquad \forall i \in \{1, 2, \ldots, k\}. \tag{3.54}$$

Afterwards, each cluster must be further examined to ensure that all its elements are continuously distributed over time, otherwise it must be further divided into more clusters, so that at last $\hat{K} \geq \hat{k}$ time-continuous CTF sets are obtained. The complete process is illustrated in Fig. 3.15.

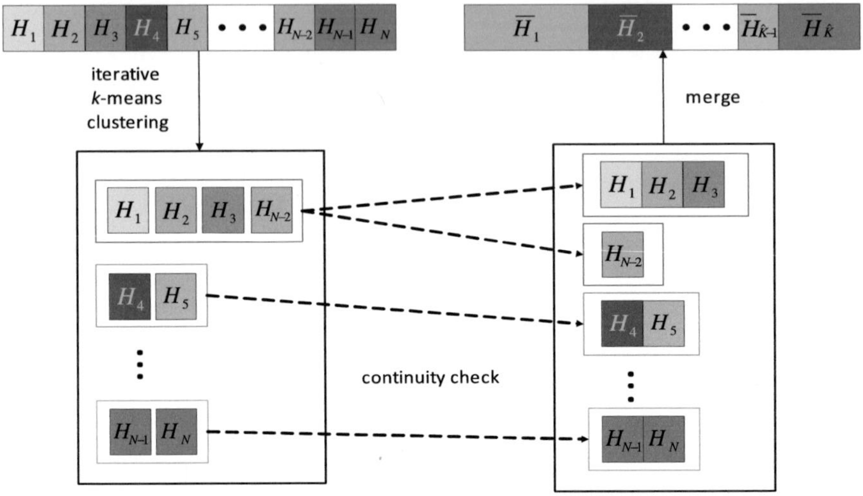

Figure 3.15 The procedure of deviation-based CTF clustering

Comparison between Two Segmentation Methods

Applying the adaptive clustering on the LPTV model shown in Fig. 3.16 with the weighting factor $w = 1$ and the threshold $\sigma_{\mathrm{max}} = 0.003$, $\hat{K} = 4$ different clusters are obtained. The LSTV approximation is shown in Fig. 3.17. Taking $K_{\mathrm{fixed}} = 4$ with the uniform segmentation method, the LSTV approximation is shown in Fig. 3.18. The deviation-based clustering shows a better performance.

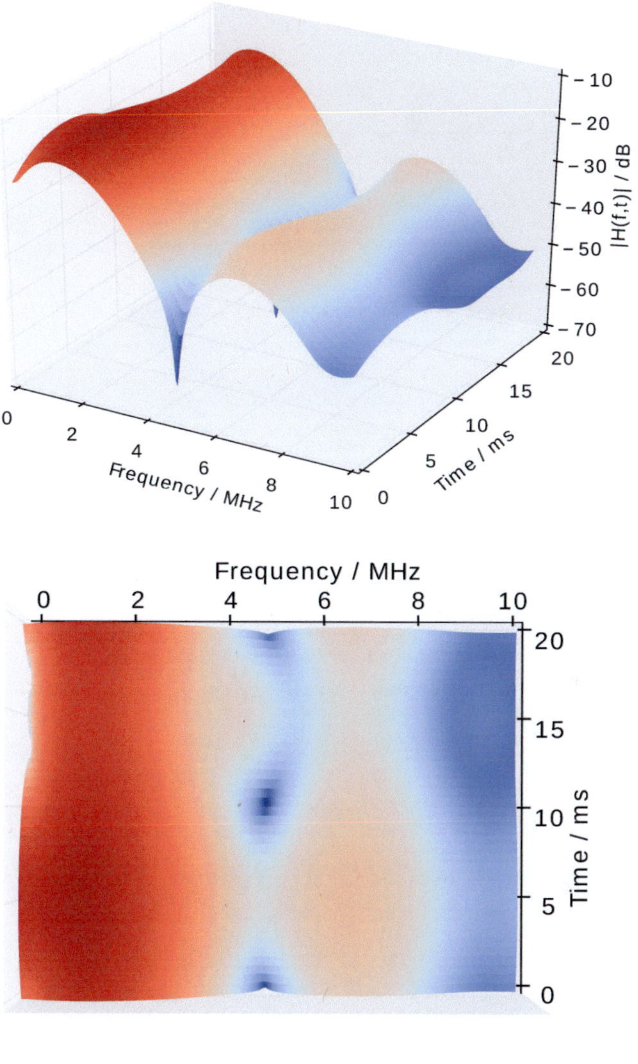

Figure 3.16 An LPTV model in the views of different angles

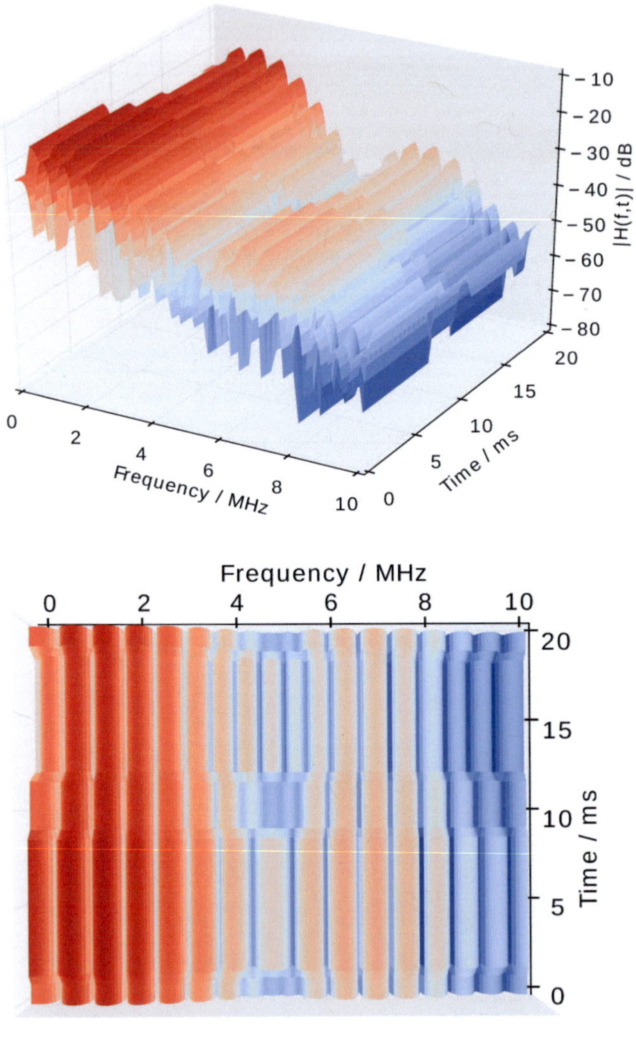

Figure 3.17 An LSTV approximation of the LPTV model from Fig.3.16, segmented with deviation-based clustering, $w = 1$, $\sigma_{\text{max}} = 0.003$

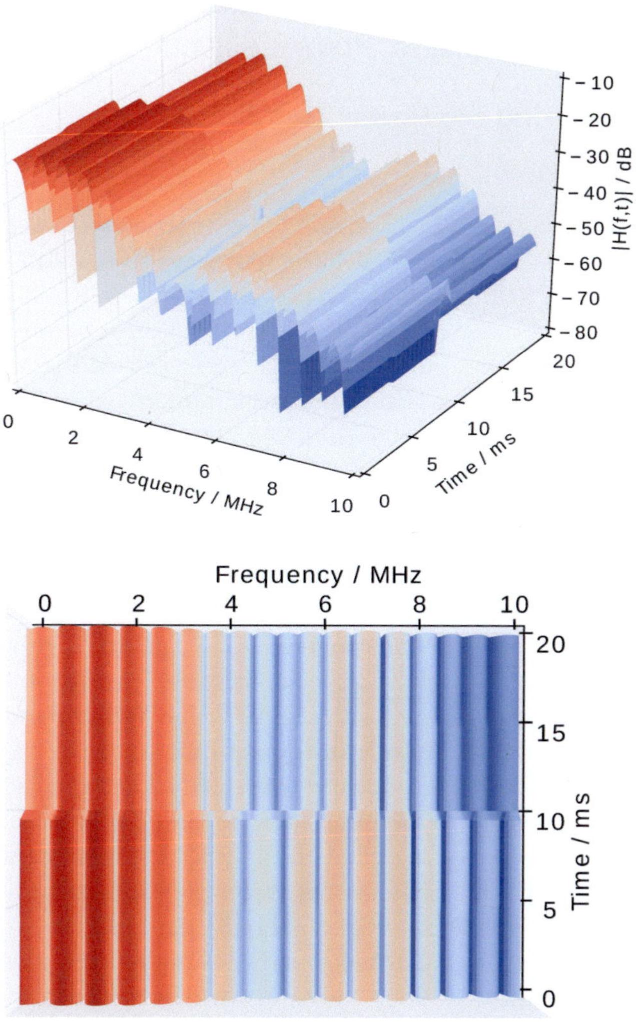

Figure 3.18 Another LSTV approximation of the LPTV model from Fig.3.16, segmented with uniform segmentation, $K_{\text{fixed}} = 4$

3.2 Measuring and Modeling of Interference Phenomena at Power Lines

3.2.1 Capturing the Power Line Noise

3.2.1.1 Mains Filter

Differing from the noise measurement in other channels, such as wireless channels, power line noise measurement faces two unique challenges. First, as referred to in Sec. 2.2.1, the power lines can transmit either in CM or in DM. Usually, the DM noise is of more interest because most PLC devices operate in DM. However, it has been reported that the CM noise can be significantly increased in the measurement, when the measuring system is directly connected to the same PE as the point under measurement [101]. Second, the power supply of the measuring system, when connected to the local power line network, generates its own noise and couples it into the channel through the cable, so that the noise scenario will be changed. Therefore, the measuring system must be EM-isolated from the point under measurement in the frequency range of interest.

Generally, the first problem can be solved by grounding the measuring system to the PE through a large capacitance, instead of a direct connection [8]. To overcome the second challenge, two different solutions are available. First, all the measuring equipment can be supplied with batteries and without PE connection [101]. Alternatively, the equipment can also be connected to the mains at a power outlet far from the point of measurement, i.e. on another phase or in a neighbor building, and through a mains filter, which consists of an isolation transformer and a line impedance stabilization network (LISN) [8]. An LISN is a well-designed low-pass filter, which passes the 50 or 60 Hz mains voltage and strongly attenuates all high-frequency signals as well as noises in the frequency range of interest. An improved design of the off-the-shelf mains filter was reported by *Liu* [55], which replaces the magnetic core coils in the LISN with air core coils, and replaces the isolation transformer connected to the mains with an uninterruptible power supply (UPS).

According to the principles proposed in [55] of designing LISNs with air coils, an LISN for the frequency range up to 10 MHz was designed as depicted in Fig. 3.19(a) with the parameters listed in Table. 3.5. Its $|S_{21}|$ parameter was measured with a VNA, as shown in Fig. 3.19(b). Strong attenuation is achieved up to 10 MHz. Connecting the LISN to a power outlet, the voltages on both sides were measured as shown in Fig. 3.19(c). It can be seen that the 50 Hz mains voltage is not distorted or attenuated by the LISN.

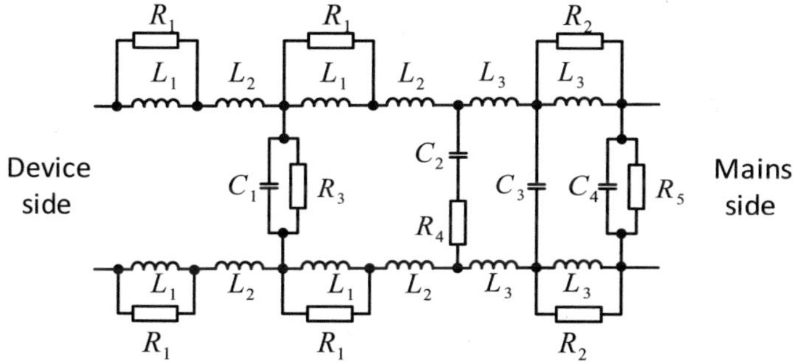

(a) Schematic diagram of the LISN

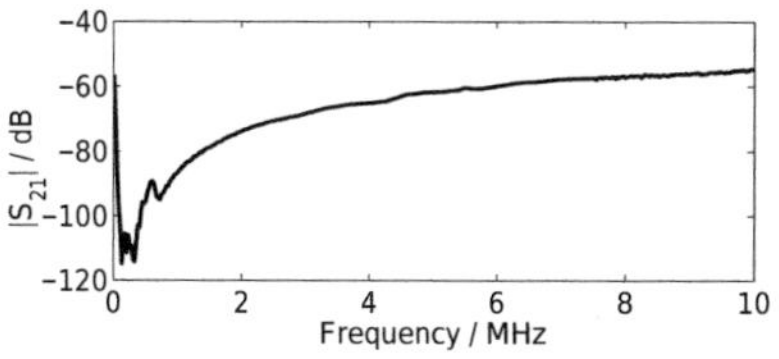

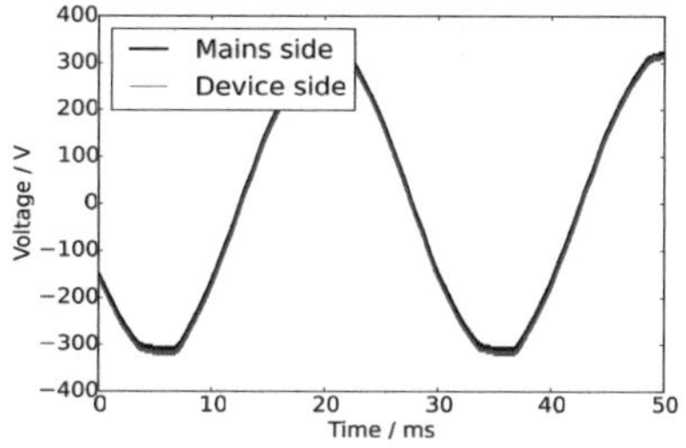

(b) The VNA-measured $|S_{21}|$ of LISN

(c) A 50 Hz mains voltage through the LISN

Figure 3.19 LISN for the frequency range up to 10 MHz

Table 3.5 Design parameters of the LISN

Inductor	L_1	L_2	L_3		
Value (µH)	20	2.8	2400		
Capacitor	C_1	C_2	C_3	C_4	
Value (nF)	100	100	330	1000	
Resistor	R_1	R_2	R_3	R_4	R_5
Value (kΩ)	0.47	0.5	1000	0.05	1000

3.2.1.2 LV Coupler Design

Similar to PLC communication systems, a PLC noise measuring system always needs a band-pass coupling stage, which simultaneously blocks the mains voltage of high amplitude, while transmitting the high-frequency noise without attenuation. Furthermore, high-voltage wide-band spikes can occur on grids at any time and damage the analog-to-digital converters (ADCs) in the measuring devices. Hence, protection circuits against such transients must be integrated into the couplers.

Basically, inductive, capacitive and hybrid structures are possible for AC signal coupling. The simplest inductive coupler is a transformer; a simple capacitive one is a pair of capacitors connecting the source and the load. With a cascade of both of them, a hybrid coupler can be implemented. For LV PLC applications, inductive couplers cannot be applied because they have negligible impedances at the mains frequency and hence shorten the LV grid. Pure capacitive couplers are not preferred either, due to their bad behavior in the low frequency range. Hence, the hybrid structure is generally chosen, especially in the transformer-capacitor design, which is briefly illustrated by Fig. 3.20. Compared to simple LC bandpass filters, this design exhibits advantages in mains isolation and spike limiting [26].

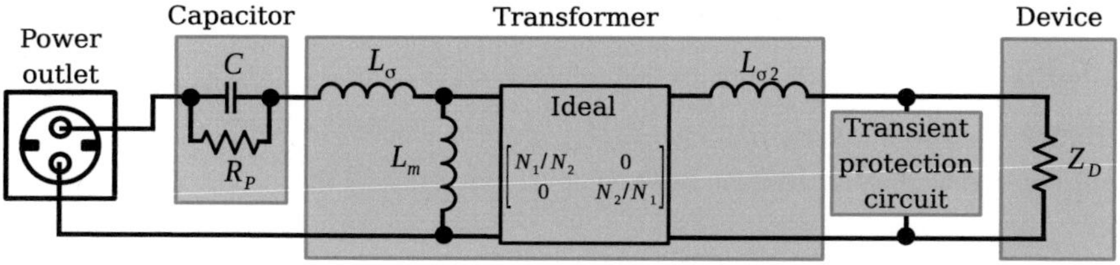

Figure 3.20 Transformer-capacitor design of LV PLC couplers: R_P is the parasitic resistance of the capacitor, L_σ and $L_{\sigma 2}$ are the leakage inductances on the primary and secondary sides of the transformer, respectively. Z_D is the equivalent impedance of the PLC device. [103]

In PLC modems, the threshold of saturation must be taken into consideration when selecting the coil core, because the transmitting power must be high enough to overcome the strong attenuation of PLC channels. Here, in the case of noise measurement, the saturation threshold is less critical, but a high impedance is required to avoid coupling loss. Hence, to implement the transformer with a large inductance and a reasonable size, magnetic zinc is selected as the material of the coil core. To optimize the coupling performance between 0.15-10 MHz, both the transformer and the capacitor are carefully optimized through simulation; the results are listed in Tab. 3.6.

Table 3.6 Optimized parameters of the LV PLC coupler for capturing noise between 0.15-10 MHz: L is the series inductance of the transformer, f_r is the resonant frequency.

C	L	N_1/N_2	$f_r = (2\pi\sqrt{LC})^{-1}$
440 nF	105 µH	1	23.4 kHz

Although common transient voltage suppressor (TVS) diodes are widely used in transient protection circuits, they are not applicable for PLC couplers working above 1 MHz, because their high capacity can seriously impact the frequency response of the couplers. Therefore, ultra low capacitance TVS arrays, Schottky diodes or even positive intrinsic negative diodes with capacities less than one picofarad are preferred [20].

3.2.1.3 Capturing the Overall Channel Noise

In the behavioral power line channel model (see Fig. 2.10), the overall channel noise $r(t)$ on the receiving side is of interest. To capture $r(t)$, the measuring system was set up as illustrated in Fig. 3.21. A universal software radio peripheral (USRP), which has two independent receiving channels with AFEs and ADCs, played the role of receiver. One channel was connected to the power outlet under test through an LV PLC coupler and a $0/20$ dB switchable attenuator, in order to sample the channel noise. When high peaks occur and cause a clipping effect, the attenuator should be switched to 20 dB for extra attenuation; otherwise it can be set to 0 dB. This attenuation must be compensated in post-processing of the raw measurement data, and it should be noted that the extra 20 dB attenuation raises the quantization error when activated, due to the fixed quantizing level. The other channel received the output of a ZCD, which was also connected to the same power outlet, so that the mains phase could be obtained. A personal computer (PC) was connected to the USRP for controlling and data storage. Both the USRP and the PC were supplied by a UPS with the aforementioned LISN to avoid influencing the channel noise. Some key specifications of the USRP are listed in Table 3.7. Note that the input impedance of the USRP is only 50 Ω, which leads to a certain level of coupling loss.

Two sample measurements are shown in Fig. 3.22, both as time waveforms and spectrograms. Different types of noises can be observed in the measurements and the higher quantization error in the measurement of office noise B appears in the form of background noise.

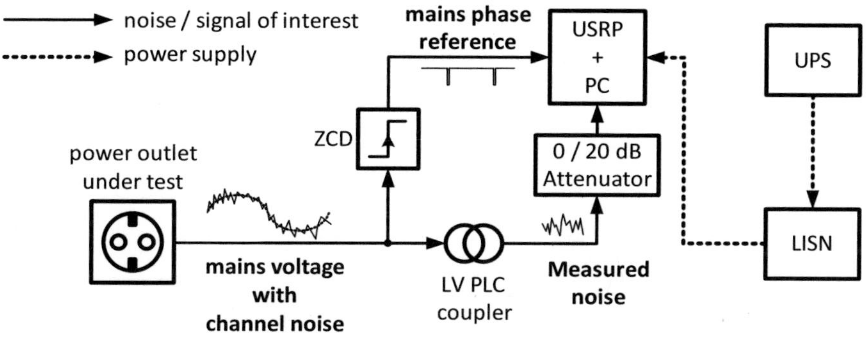

Figure 3.21 Hardware setup for capturing overall channel noise

Table 3.7 Key specifications of the USRP

Sampling rate	25 MSPS
Input range	$\pm 1.25\,\mathrm{V}$
Resolution	14 bits
Input impedance	$50\,\Omega$

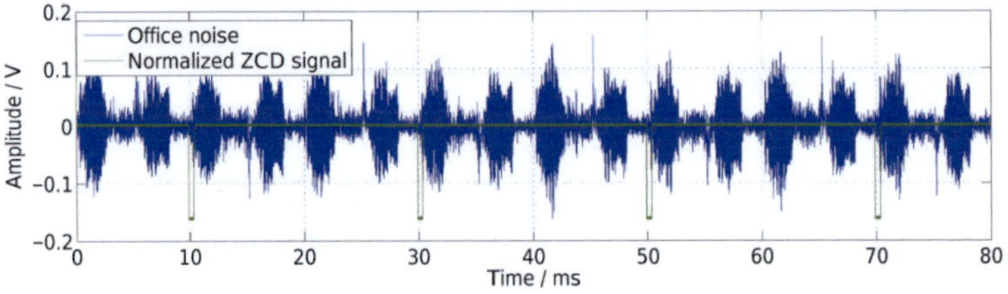

(a) Office noise A: time waveform

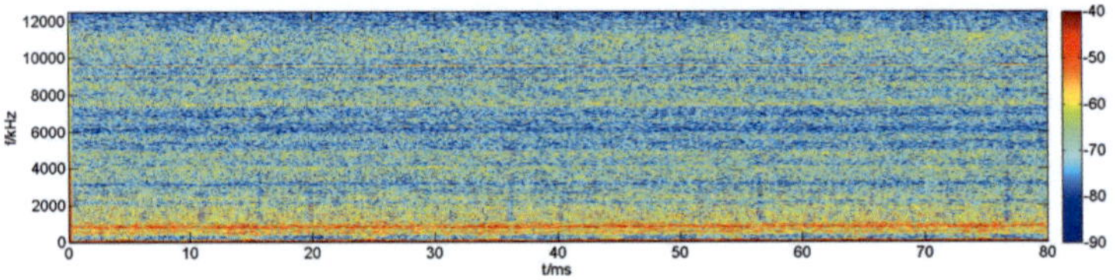

(b) Office noise A: spectrogram

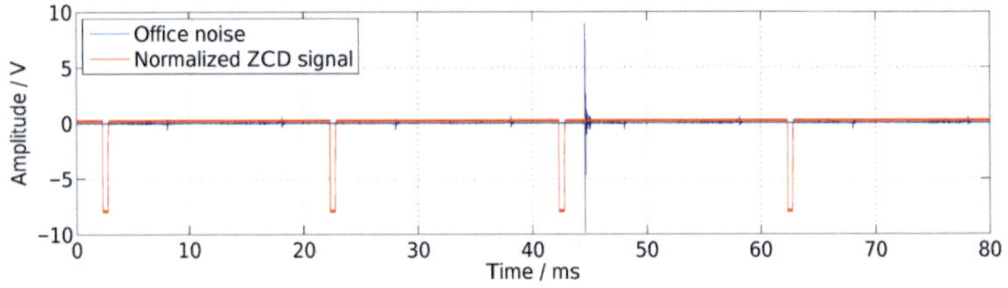

(c) Office noise B: time waveform

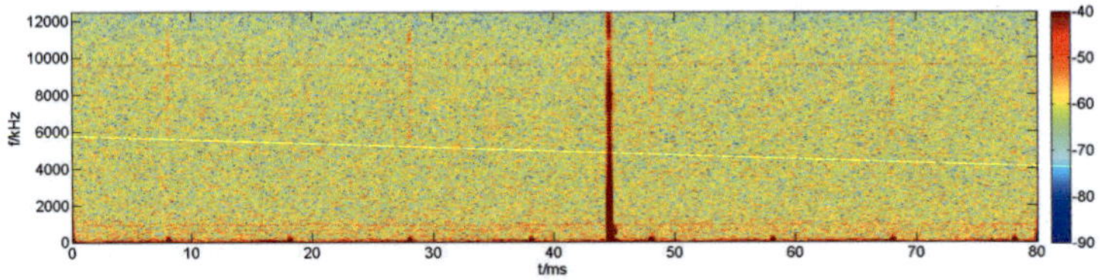

(d) Office noise B: spectrogram

Figure 3.22 Two sample channel noise measurements captured at a power outlet in the laboratory: Note that the extra 20 dB attenuation was deactivated when measuring office noise A, and activated when measuring office noise B.

3.2.1.4 Capturing the Noise Generated by a Power Consumer

Although the overall channel noise is of major interest in the behavioral model, it is also worth to discuss about capturing the noise generated by a power consumer. As shown in Fig. 3.23, the consumer under test is supplied by a UPS with LISN instead of a regular power outlet, so that no extra power line noise but only the noise generated by the consumer is captured. Two sample measurements are given in Fig. 3.24. It can be seen that the noise pattern highly depends on the type of consumer: the dimmer lamp generates almost only periodical impulses, while the monitor is generating a significant broadband background noise and some narrowband noise besides periodical impulses.

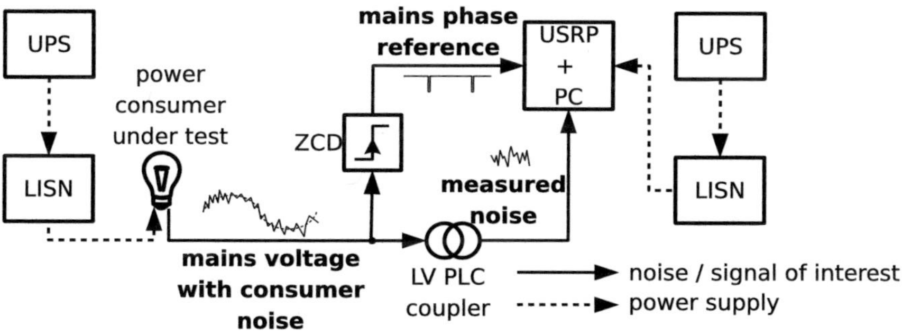

Figure 3.23 Hardware setup for capturing consumer noise

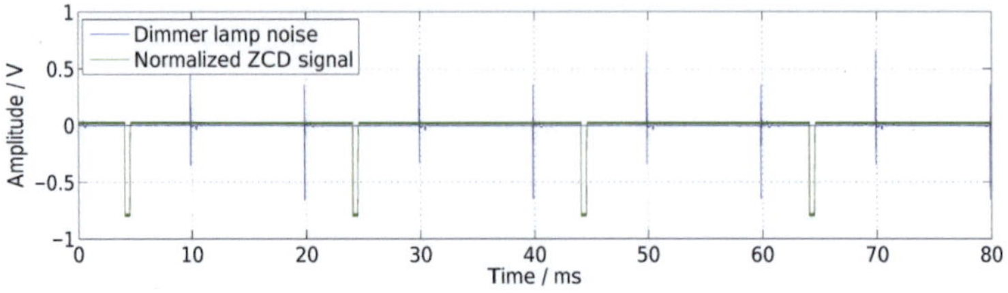

(a) Dimmer lamp noise: time waveform

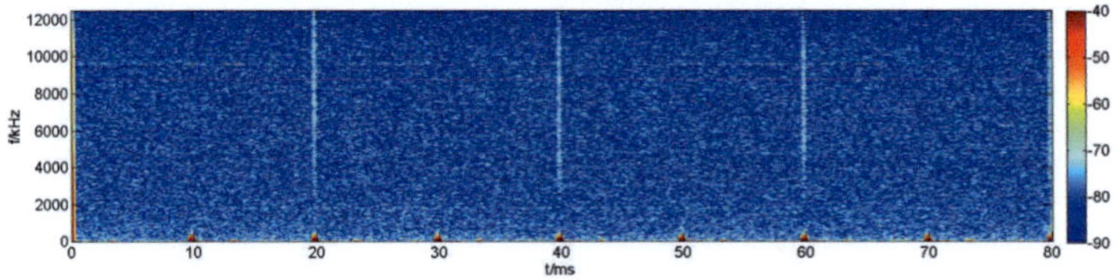

(b) Dimmer lamp noise: spectrogram

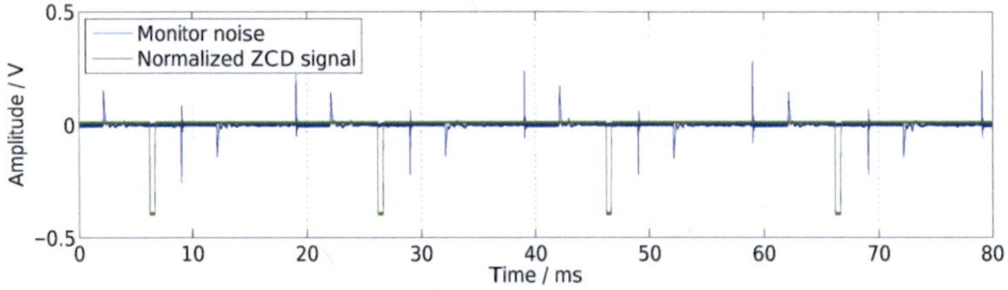

(c) Monitor noise: time waveform

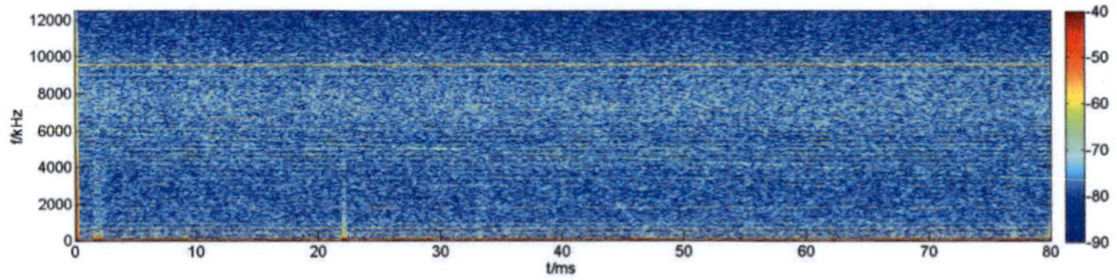

(d) Monitor noise: spectrogram

Figure 3.24 Noises generated by a dimmer lamp and a monitor

3.2.2 Additive Power Line Channel Noise Model

As introduced in Sec 2.3.3 and shown with measurements in Sec. 3.2.1.3, the overall power line channel noise is always a mixture of a variety of interferences, which behave differently in the time-frequency domains. These interferences are generally considered to be generated by different sources, and are thus independent of each other, but received together at the output side of power line channels. According to this idea, the power line noise was first classified for narrowband applications by *Hooijen* [42], and then extended for broadband applications by *Zimmermann* et al. [98]. Since then, it has been common to model the power line noise scenarios with the additive model

$$n(t) = n_{\mathrm{NBN}}(t) + n_{\mathrm{PINS}}(t) + n_{\mathrm{PINAS}}(t) + n_{\mathrm{APIN}}(t) + n_{\mathrm{CBGN}}(t), \quad (3.55)$$

where n is the overall channel noise at the receiver, as in Fig. 2.10. n_{NBN},n_{PINS},n_{PINAS}, n_{APIN} and n_{CBGN} are the components of narrowband noise, periodic impulsive noise synchronous to the mains voltage, periodic impulsive noise asynchronous to the mains voltage, aperiodic impulsive noise and colored background noise, respectively. As aforementioned in Sec. 2.3.3, it is also usual to call PINS and PINAS collectively as periodic impulsive noise, so that the model can be written as

$$n(t) = n_{\mathrm{NBN}}(t) + n_{\mathrm{PIN}}(t) + n_{\mathrm{APIN}}(t) + n_{\mathrm{CBGN}}(t), \quad (3.56)$$

where $n_{\mathrm{PIN}} = n_{\mathrm{PINS}} + n_{\mathrm{PINAS}}$ is the component of periodic impulsive noise. For a deeper noise analysis, each of these noises must be individually modeled.

3.2.3 Modeling and Estimating Narrowband Noise

3.2.3.1 NBN Model

The narrowband noise, as introduced in Sec. 2.3.3, is mainly caused by electrical appliances and radio signals. Usually in an interference scenario, multiple narrowband interferers of different central frequencies, bandwidths and powers exist at the same time. So the NBN component can be described by a linear model in the form

$$n_{\text{NBN}}(t) = \sum_{i=1}^{N_{\text{NBN}}} c_{\text{NBN},i}(t) = \sum_{i=1}^{N_{\text{NBN}}} A_{\text{NBN},i}(t) sin(2\pi f_{\text{C},i}t + \varphi_i(t)), \quad (3.57)$$

where N_{NBN} is the number of narrowband interferers, $A_{\text{NBN},i}(t)$, $f_{\text{C},i}$ and $\varphi_i(t)$ are the time envelope, the central frequency and the phase offset of the ith narrowband interferer $c_{\text{NBN},i}(t)$, respectively.

For each individual interferer $c_{\text{NBN},i}(t)$ received at the output of the channel, the time envelope $A_{\text{NBN},i}(t)$ can origin in two kinds of sources: the noise source and the channel. Depending on its nature, the interferer can have an envelope at its source that is either constant (e.g. a frequency modulated (FM) radio signal), wide-sense stationary (e.g. an AM radio signal) or cyclostationary (e.g. when generated by an electrical appliance with periodical characteristics). The channel, due to its cyclostationarity with respect to the access impedance and the CTF, usually contributes to a periodicity in the time envelope. Depending on the sort of the noise source and the strength of cyclostationarity of the channel, $c_{\text{NBN},i}(t)$ can be either wide-sense stationary or cyclostationary, as reported by *Cortés*[16] and *Opalko*[69]. Recalling the conclusion in Sec. 2.3.1 that all wide-sense stationary processes are also cyclostationary, it can be generally asserted that $A_{\text{NBN},i}(t)$ is always cyclostationary.

For a simplification, the time envelope can be further decomposed into two terms

$$A_{\text{NBN},i}(t) = A_{\text{D},i}(t)A_{\text{S},i}(t), \quad (3.58)$$

where the deterministic term $A_{\text{D},i}(t)$ is either a constant or a periodical function of time, while the stochastic term $A_{\text{S},i}(t)$ is a wide-sense stationary process over time. Furthermore, considering the nature of NBN sources and the low-pass character of CTF, $A_{\text{S},i}(t)$ is always limited in bandwidth and amplitude. So that (3.57) turns into

$$\begin{aligned} n_{\text{NBN}}(t) &= \sum_{i=1}^{N_{\text{NBN}}} A_{\text{D},i}(t) A_{\text{S},i}(t) sin(2\pi f_{\text{C},i} t + \varphi_i(t)) \\ &= \sum_{i=1}^{N_{\text{NBN}}} A_{\text{D},i}(t) sin(2\pi t(f_{\text{C},i} + \Delta f_i(t))), \end{aligned} \tag{3.59}$$

where

$$|\Delta f_i(t)| \leq \frac{B_i}{2}, \tag{3.60}$$

and B_i is the bandwidth of $c_{\text{NBN},i}$. Thus, every individual narrowband interferer can be characterized with its central frequency, bandwidth and deterministic envelope.

3.2.3.2 Estimating Narrowband Noise Using the Modified Welch's Method

As discussed above, to characterize the NBN component in a power line noise measurement, every individual narrowband interferer should be described with its central frequency, bandwidth and deterministic envelope. Furthermore, it is essential to cancel the NBN from the noise, in case that it is influencing the analysis of noise of other types. *Liu* has proposed a NBN extraction algorithm in [55], which cancels a narrowband interferer with given central frequency and bandwidth, and estimates its deterministic envelope. Therefore, the task of NBN characterization can be simplified to a narrowband interferer detection and a bandwidth estimation. To achieve this aim, a spectral estimation must be carried out with the noise measurement.

A simple non-parametric estimator is the periodogram

$$\hat{S}[k] = \hat{S}(k\Delta f) = \frac{T_{\text{S}}}{N} |\sum_{n=0}^{N-1} x[n] w[n] e^{-j2\pi nk\Delta f}|^2, \tag{3.61}$$

where $x[n] = x(nT_{\text{S}})$ is the sampled measurement, $w[n]$ is window function, T_0 is the sampling interval, N is the number of FFT points and Δf is the frequency resolution. However, for power line noises, the accuracy of periodogram can be significantly reduced by the presence of large impulses, which raise the overall noise power in a wide frequency range. As an example, different types of noise were synthesized as shown in Fig. 3.25. The CBGN was generated by filtering an AWGN with an LTI low-pass filter; the NBN was generated after the model (3.59) with only one interferer, by which $f_C = 100\,\mathrm{kHz}$, $B = 3\,\mathrm{kHz}$ and the envelope was a 50 Hz triangle wave; the PIN is 100 Hz cyclic. First, the noise was constructed with the CBGN and the NBN. Afterwards, the PIN was also added in. In both cases the periodograms were computed with a Hamming window, as shown in Fig. 3.26. It can be seen that the PIN reduced the significance of the NBN peak, and even created several new peaks in the periodogram.

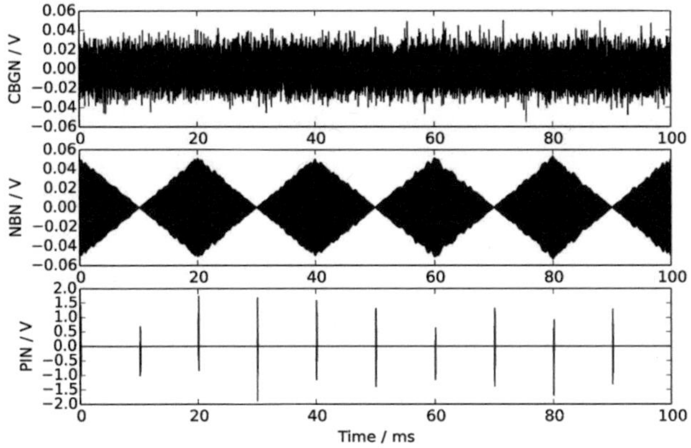

Figure 3.25 Noise components, simulated with a synthesizing rate of 1 MHz

To mitigate this effect, the Welch's method is usually applied, i.e. the measurement is divided into several segments of the same length, the periodogram of each segment is computed and the averaged periodogram is taken as the spectrum estimation. This method can be represented as

$$\hat{S}_{\text{Welch}}[k] = \frac{T_{\text{S}}}{ML} \sum_{m=0}^{M-1} \left| \sum_{l=0}^{L-1} x[mL_{\text{step}} + l]w[l]e^{-j2\pi lk\Delta f} \right|^2, \tag{3.62}$$

where L is the window length, L_{step} is the step length, $M = \lfloor \frac{N}{L_{\text{step}}} \rfloor$ is the number of segments and $1 - \frac{L_{\text{step}}}{L}$ is the overlapping rate.

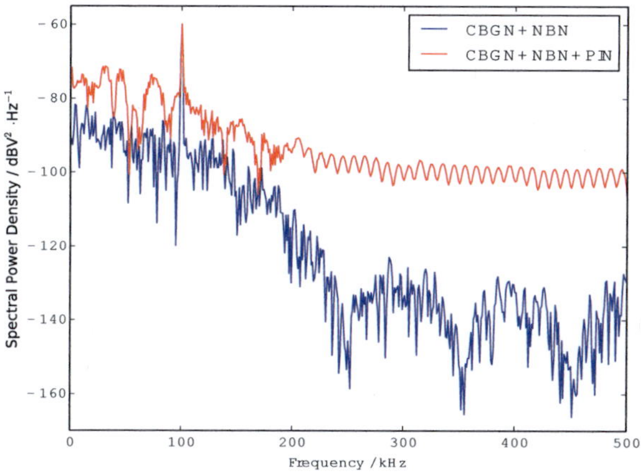

Figure 3.26 Periodograms of noises with and without a PIN component

Considering the impact of high-valued impulses on the averaging process, this method can be further improved by using the median of the periodograms instead of the average. The result of the improved method are depicted in Fig. 3.27. It can be seen that the modified Welch's method shows the best performance.

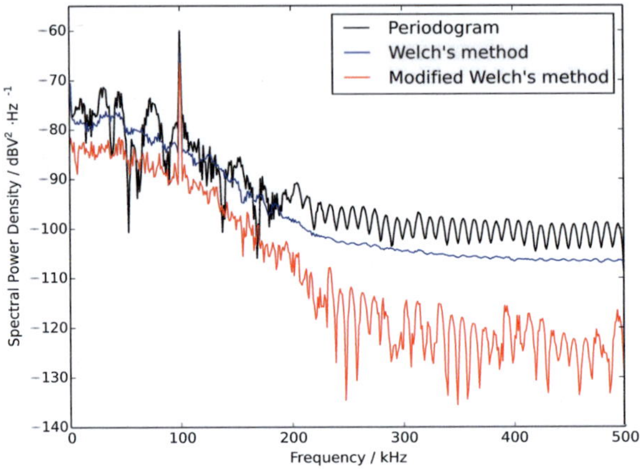

Figure 3.27 Improved spectral estimators, compared to the periodogram: The Hamming window is applied in all methods. The overlapping rate is 0 in the Welch's method and the modified Welch's method.

By detecting peaks in the estimated spectrum, the central frequencies of the existing NBNs are obtained. As a next step, the bandwidths are to be investigated. Some NBNs are generated by licensed radio stations, whose spectral allocations are stable and publicly available. The bandwidths of these NBNs can be conveniently checked, according to the central frequencies. However, the other NBNs, e.g. the noises generated by electrical appliances and amateur radio signals, can have flexible and unknown bandwidths. Therefore, the bandwidth must be estimated. A simple solution is using an attenuation threshold leading to

$$\hat{f}_{\mathrm{u},i} = \arg\max_{f_\mathrm{u}}\{S(f) \leq S(f_{\mathrm{C},i}) - \gamma, \forall f \in [f_{\mathrm{C},i}, f_\mathrm{u}]\}, \tag{3.63}$$

$$\hat{f}_{\mathrm{l},i} = \arg\min_{f_\mathrm{l}}\{S(f) \leq S(f_{\mathrm{C},i}) - \gamma, \forall f \in [f_\mathrm{l}, f_{\mathrm{C},i}]\}, \tag{3.64}$$

$$\hat{B}_i = \hat{f}_{\mathrm{u},i} - \hat{f}_{\mathrm{l},i}, \tag{3.65}$$

where γ is the attenuation threshold in dB.

3.2.4 Modeling and Estimating Aperiodic Impulsive Noise

3.2.4.1 Aperiodic Impulsive Noise Model

The aperiodic impulsive noise, as introduced in Sec. 2.3.3, can be generated by operations of connection and disconnection of electrical devices in the network. The APIN is generally broadband, has no deterministic pattern of occurrence or waveform, and thus can be described only by stochastic models.

Existing APIN models can be generally divided into two classes. The first class models the APIN together with the background noise and focuses on the probability distribution of the noise amplitude, such as Middleton's Class-A model [26, 62]

$$p(n) = \sum_{m=0}^{+\infty} P_m \frac{1}{\sqrt{2\pi\sigma_m^2}} e^{-\frac{n^2}{2\sigma_m^2}}, \tag{3.66}$$

$$P_m = \frac{e^{-A}A^m}{m!}, \tag{3.67}$$

$$\sigma_m^2 = \sigma^2 \frac{m/A + \Gamma}{1+\Gamma} = \sigma_i^2 \frac{m}{A} + \sigma_g^2, \tag{3.68}$$

where $p(n)$ is the PDF of the noise voltage n, σ_g^2 and σ_i^2 are the powers of the Gaussian (background) noise and the impulsive noise, respectively, and $\Gamma = \sigma_g^2/\sigma_i^2$ is the mean ratio between them. A is the so-called impulsive index, which is smaller when the noise is more impulsive, and approaches $+\infty$ when the noise is Gaussian. The second class models the APIN with two independent aspects: the occurrence and the probability distribution. Two well-known models in this class are the Bernoulli-Gaussian model [35, 73] and the partitioned Markov chain model [100]. Generally both of them can be represented in a time-discrete form as

$$n_{\text{APIN}}[k] = \Phi[k] n_{\text{G}}[k], \tag{3.69}$$

where n_{G} is Gaussian noise and $\Phi \in \{0, 1\}$ is a random process that denotes the occurrence of APIN. With APIN it takes on the value of 1 in absence of APIN it is 0. In the Bernoulli-Gaussian model, Φ is a Bernoulli process, which is relatively simple. In the partitioned Markov chain model Φ is described with the Markov process shown in Fig. 3.28.

In this model, N_D impulse-free states, N_W impulsive states and two transition states are defined. The transition states are only intented to simplify the model and have no physical meaning. The probabilities of transitions between states can be summarized with the matrix

$$\mathbf{P} = \begin{bmatrix} u_{1,1} & \cdots & 0 & u_{1,t_1}g_{t_1,1} & \cdots & u_{1,t_1}g_{t_1,N_W} \\ \vdots & \ddots & \vdots & \vdots & \ddots & \vdots \\ 0 & \cdots & u_{N_D,N_D} & u_{N_D,t_1}g_{t_1,1} & \cdots & u_{N_D,t_1}g_{t_1,N_W} \\ g_{1,t_2}g_{1,t_2} & \cdots & g_{1,t_2}u_{t_2,N_D} & g_{1,1} & \cdots & 0 \\ \vdots & \ddots & \vdots & \vdots & \ddots & \vdots \\ g_{N_W,t_2}u_{t_2,1} & \cdots & g_{N_W,t_2}u_{t_2,N_D} & 0 & \cdots & g_{N_W,N_W} \end{bmatrix}, \quad (3.70)$$

where

$$u_{i,t_1} = 1 - u_{i,i}, \quad (3.71)$$

$$g_{i,t_2} = 1 - g_{i,i}. \quad (3.72)$$

Grouping the impulse-free states into the set $\mathbf{A}$ and the impulsive states into $\mathbf{B}$, the occurrence process Φ is given by the Markov chain as

$$\Phi[k] = \begin{cases} 0, & z[k] \in \mathbf{A} \\ 1, & z[k] \in \mathbf{B} \end{cases}, \quad (3.73)$$

where $z[k]$ is the state at the kth sample.

Comparing with the first class of APIN models, the second class shows three advantages. First, the APIN is separated from the background noise. Second, with the occurrence process Φ, not only the occurrences, but also the widths of impulses can be quantitatively modeled. At last, by cascading two layers of the occurrence process, the burst events can be modeled describing the phenomenon that several APIN impulses arrive during a short time. Therefore, the partitioned Markov chain model is taken here for APIN. For a further simplification (3.69) is modified into

$$n_{\text{APIN}}[k] = \sigma^2_{\text{APIN}}[k]\Phi[k]n_{\text{W,Uni}}[k], \tag{3.74}$$

where $n_{\text{W,Uni}}$ is a zero-mean white Gaussian noise with unity power and σ_{APIN} is the variance, which is random for every individual impulse. It remains invariable through one impulse, and is zero when the APIN is absent. For example, in a measurement with N aperiodic impulses, such as shown in Fig. 3.29, there is

$$\sigma_{\text{APIN}}[k] = \begin{cases} \sigma_{\text{APIN},i}, & \frac{t_{\text{APIN},i}}{T_{\text{S}}} \le k \le \frac{t_{\text{APIN},i}+\tau_{\text{APIN},i}}{T_{\text{S}}}, i = 1, 2, \ldots, N \\ 0, & \text{otherwise}, \end{cases}, \tag{3.75}$$

where T_{S} is the sampling interval.

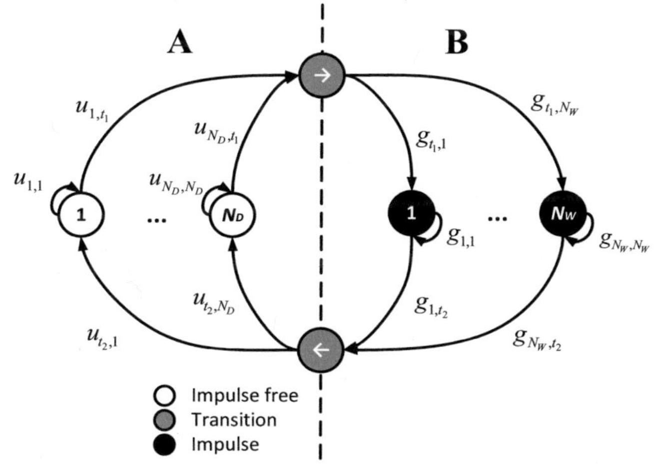

Figure 3.28 Partitioned Markov chain model, representing APIN occurrence $u_{i,j}$ and $g_{i,j}$ represent the probabilities of transitions between two states.

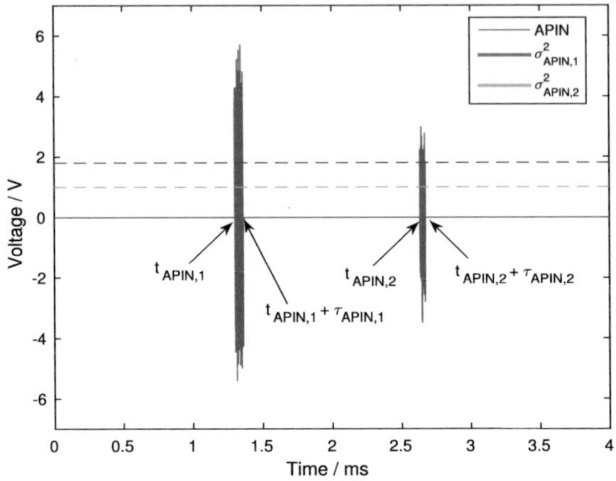

Figure 3.29 Two aperiodic impulses and their key parameters: $\sigma^2_{\mathrm{APIN},i}$,$t_{\mathrm{APIN},i}$ and $\tau_{\mathrm{APIN},i}$ are the variance, the arriving time and the width of the ith impulse, respectively.

3.2.4.2 Extracting Aperiodic Impulses Using the Empirical Mode Decomposition

With a given noise measurement, the probability matrix $\mathbf{P}$, and thus also the occurrence process Φ can be estimated, based on the statistics of inter-arrival time and widths of the impulses, using the method proposed by *Liu* in [55]. The inter-arrival time (IAT) between the ith and the $(i+1)^{\mathrm{th}}$ impulse can be simply computed by subtracting their arriving times

$$\Delta t_{\mathrm{APIN},i} = t_{\mathrm{APIN},i+1} - t_{\mathrm{APIN},i}. \tag{3.76}$$

The amplitude factor can be modeled with a Gaussian random process, according to its distribution. Thus, the estimation of APIN equals the estimations of the arriving time, the width and the amplitude of every aperiodic impulse. In most work referring to the topic of APIN analysis, the impulses are only detected but not extracted [16, 100], because the high power of aperiodic impulses enables the estimations of their amplitudes and widths, even when they are mixed with low-powered noise of other types. However, the existence of high-powered APIN can potentially degrade the analysis of other noises such as PIN and CBGN, as it will be discussed in Sec. 3.2.5.2. A reliable APIN extraction method is therefore required.

Liu has reported a time-frequency domain approach which detects and removes impulses from a noise measurement in [55], but regardless of the noise type, i.e. PINs and even NBNs with impulsive envelopes can also be removed together, so that further processing is still needed. Another solution has been proposed by *Lin* et al. which makes use of the high time sparsity of APIN. This solution estimates the APIN with help of sparse Bayesian learning (SBL). However, the SBL technology depends on compressive sensing, which requires specialized hardware, such as nonuniform samplers [22], and thus does not match the measuring setup in this work. Hence, we developed a new APIN extraction algorithm, based on the empirical mode decomposition (EMD) and median filtering.

Empirical Mode Decomposition

The EMD, also known as part of the Hilbert-Huang transform (HHT), was proposed by *Huang* et al. in [43] and has proved to be a powerful time domain method for analyzing non-stationary signals. Comparing to the other analyzing methods for non-stationary signals, such as the short-time Fourier tansform (STFT) [38] and the locally stationary wavelet (LSW) analysis [64], the EMD does not leave the time domain and performs independently of any selection of a window or wavelet bases. It decomposes a given time signal into a set of intrinsic mode functions (IMFs), which are defined as functions satisfying the following two conditions:

1. *in the whole data set, the number of extrema and the number of zero-crossings must either be equal or differ by one;*
2. *at any point, the mean value of the envelope, defined by the local maxima and the envelope defined by the local minima is zero.*

This definition forces an IMF to be wide-sense stationary, or at least locally stationary, which was derived in detail in [43].

Based on the concept of IMF, for any data series $x(n)$ with two or more extrema, the EMD method can be defined as follows:

1. Identify all local extrema. Connect all local maxima with a cubic spline line as the upper envelope $u(n)$, then connect all local minima for the lower envelope $l(n)$. Define m_1 as the mean of the two envelope:

$$m_1(n) = \frac{u(n) + l(n)}{2}. \tag{3.77}$$

2. Subtract m_1 from x, use the result to update the original input, and repeat the algorithm iteratively:

$$h_1(n) = x(n) - m_1(n), \tag{3.78}$$

$$m_{11}(n) = \frac{u_1(n) + l_1(n)}{2}, \tag{3.79}$$

$$h_{11}(n) = h_1(n) - m_{11}(n), \tag{3.80}$$

$$h_{1k}(n) = h_{1(k-1)}(n) - m_{1k}(n), \tag{3.81}$$

   where u_1 and l_1 are the upper and lower envelopes, extracted from h_1; m_{1k} is the mean of the upper and lower envelopes, extracted from $h_{1(k-1)}$ for any $k > 1$. The iteration stops when m_{1k} falls below a given threshold, and the final h_{1k} is considered to be an IMF component

$$c_1(n) = h_{1k}(n). \tag{3.82}$$

3. Subtract c_1 from x and repeat the steps above iteratively on the residual

$$r_i(n) = \begin{cases} x(n) - c_1(n) & i = 1 \\ r_{i-1}(n) - c_i(n) & i > 1. \end{cases} \tag{3.83}$$

   The iteration will be stopped by any of the following criteria:

a) When c_k or r_k becomes smaller than a predefined threshold;
b) when the residual r_k becomes monotonic.

In the end, the original signal will be decomposed into a set of IMFs $\{c_1, \ldots, c_K\}$ and a final residual r_K, which is normally a constant offset or a trend of x with

$$x(n) = \sum_{i=1}^{K} c_i(n) + r_K(n). \tag{3.84}$$

As an example, a segment of the noise measurement in Fig. 3.22(c) with APIN is processed with the EMD and the results are depicted in Fig. 3.30. The IMFs with higher orders contain components with lower frequencies.

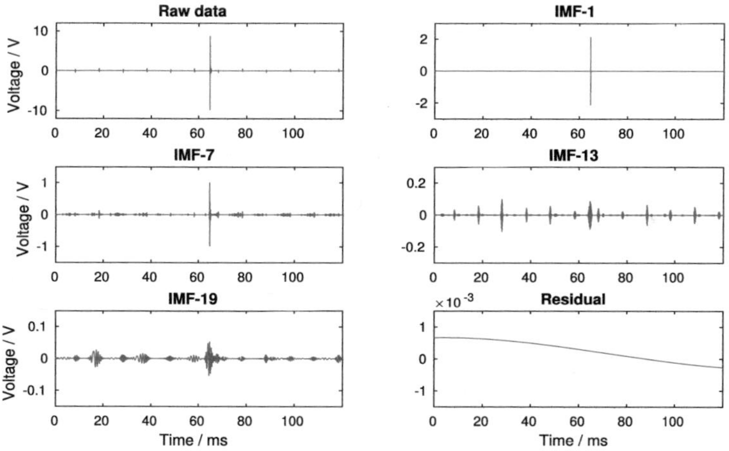

Figure 3.30 Applying the EMD on a noise measurement with APIN, 27 IMFs in total are extracted, 4 are depicted here. Note that the y-axis is differently normalized in every sub-plot.

EMD-Based APIN Detection and Extraction

Against interference in non-stationary signals, several different EMD-based artifact-removing algorithms have been developed in the recent years [25, 102, 53, 95]. Out of these interferences, the impulsive artifacts in intracranial pressure signals investigated in [25] and [102] exhibit a high similarity with the APIN at power lines by being

- random,
- aperiodic,
- impulsive (short in duration, high-powered, wide-band),
- and mainly caused by transients in circuits.

Aiming at these artifacts, *Feng* pointed out in [25] that large magnitude oscillations in the 1st IMF component can perfectly align with all the artifact episodes in the original measurement. The same phenomenon can also be observed with power line noise, as shown in Fig. 3.30. Furthermore, comparing the 1st IMFs of two segments in a same noise measurement, one with PIN and APIN components and the other with no APIN, it can be observed that the 1st IMF is insensitive to PIN, as shown in Fig. 3.31.

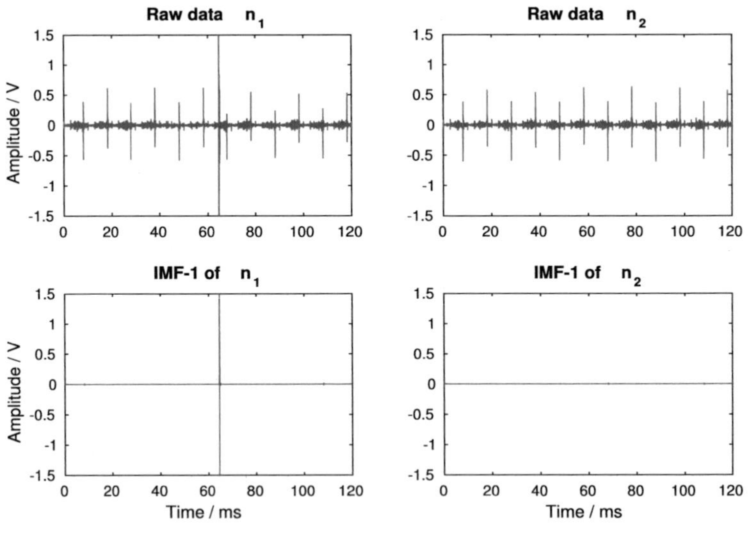

Figure 3.31 Two segments of the same noise measurement with/without APIN and their 1th IMFs

Based on the 1st IMF's sensitivity to APIN and its robustness to PIN, an APIN detector is designed as

$$D[k] = \begin{cases} 1 & |i_{n,1}[k]| \geq \mu_1 \hat{\sigma}_{i_{n,1}}, \\ 0 & \text{otherwise}, \end{cases} \tag{3.85}$$

where $i_{n,1}$ is the 1st IMF of the noise measurement n, μ_1 is a significance factor, $\hat{\sigma}_{i_{n,1}}$ is the median absolute deviation of $i_{n,1}$, denoted as

$$\hat{\sigma}_x = 1.4826 \times \text{median}_i(|x[i] - \text{median}_j(x[j])|). \tag{3.86}$$

If D remains 0 over the whole measurement, it is considered as APIN-free. Otherwise, the strongest peak is located as

$$k_{\text{peak}} = \arg \max_{k:D[k]=1} |n[k]|, \tag{3.87}$$

and a short segment of the measurement around $k = k_{\text{peak}}$ is taken for further filtering to lower the computational complexity, i.e.

$$\tilde{n}[k] = n[k + k_0], \qquad k \in [0, L_{\text{EMD}}), \tag{3.88}$$

$$k_0 = \max\{0, k_{\text{peak}} - \frac{L_{\text{EMD}}}{2}\}, \tag{3.89}$$

where L_{EMD} is the length of the segment. Due to the fact that the APIN can be usually higher by several tens of dBs than the BGN, $\mu_1 = 40$ is recommended. According to the statistics reported in [100], impulses in power line noise can last up to hundreds of µs. Therefore, depending on the sampling rate, a value of L_{EMD} corresponding to at least 1 ms is recommended. An example is shown in Fig. 3.32.

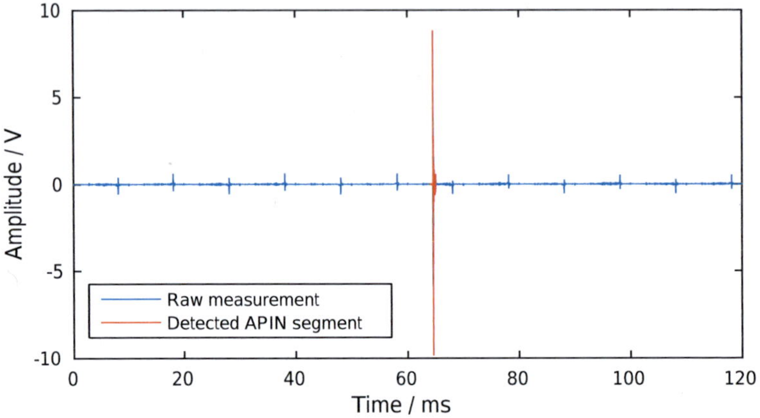

Figure 3.32 Extracting a short segment around the detected APIN from a noise measurement

Having the segment with the detected aperiodic impulse, an iterative APIN extraction is executed in the IMF domain and based on median filtering by the following steps

1. Buffer the original segment $\tilde{n}$ as $\tilde{n}_{\text{origin}}$.
2. The segment $\tilde{n}$ is decomposed into its IMFs $i_{\tilde{n},1}, \dots i_{\tilde{n},M}$ and the residual $r_{\tilde{n}}$ via the EMD.
3. Peaks in each IMF $i_{\tilde{n},j}$ are detected and smoothed by a median filter for a "cleaned" IMF

$$s_{\tilde{n},j}[k] = \begin{cases} i_{\tilde{n},j}[k] & |i_{\tilde{n},j}[k]| < \mu_2 \hat{\sigma}_{i_{\tilde{n},j}}; \\ \text{median}\{i_{\tilde{n},j}[k-L_\text{m}], \dots, i_{\tilde{n},j}[k+L_\text{m}])\} & \text{otherwise,} \end{cases} \quad (3.90)$$

where μ_2 is a significance factor of peaks, and $2L_\text{m}+1$ is the order of the median filter.

4. Summing all cleaned IMFs and the residual up to obtain the smoothed segment

$$\tilde{n}_{\text{smoothed}}[k] = \sum_{j=1}^{M} s_{\tilde{n},j}[k] + r_{\tilde{n}}[k] \tag{3.91}$$

5. Compute the error rate

$$\epsilon = \frac{\sum_{k=0}^{L_{\text{EMD}}-1} \left(\tilde{n}[k] - \tilde{n}_{\text{smoothed}}[k]\right)^2}{\sum_{k=0}^{L_{\text{EMD}}-1} \left(\tilde{n}[k] + \tilde{n}_{\text{smoothed}}[k]\right)^2/4}, \tag{3.92}$$

   if ϵ is below a threshold for convergence ϵ_0, the APIN component is extracted as $\tilde{n}_{\text{origin}} - \tilde{n}_{\text{smoothed}}$. Otherwise, update $\tilde{n}$ with $\tilde{n}_{\text{smoothed}}$ and repeat the process from step 2 on.

For example, the detected APIN in Fig. 3.32 is extracted from the measurement as shown in Fig. 3.33. The proposed method exhibits a good performance.

Iteratively applying the proposed APIN detection and extraction methods on measured noise, until no more APIN is detected, all aperiodic impulses can be separately extracted, so that the amplitude, impulse width and arrival time of each impulse can be obtained. Based on the statistics, the parameters of the partitioned Markov chain model can be estimated with the method given by *Liu* in [55].

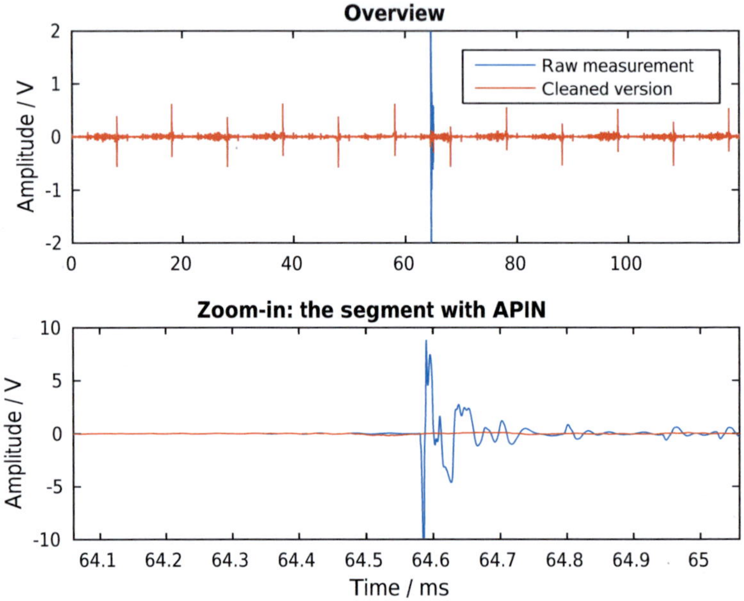

Figure 3.33 Extracting the APIN component from the measured noise with the proposed method, L_{EMD} is set corresponding to 0.5 ms with the sampling rate, $\mu_2 = 3$ and $\epsilon_0 = 1 \times 10^{-4}$

3.2.5 Modeling and Estimating Periodic Impulsive Noise

3.2.5.1 Periodic Impulsive Noise Model

The periodic impulsive noise, as introduced in Sec. 2.3.2, originates from non-linear circuits, which have periodically time-varying electrical properties, such as rectifier diodes or switching regulators. Depending on the operation period of its source device, a PIN can be either synchronous or asynchronous to the mains voltage, and be classified as PINS or PINAS, respectively. Despite of the different cyclic periods, PINS and PINAS show no difference in their behavioral pattern: periodically occurring with high amplitude and lasting for short duration. Therefore, we generally model PIN in the following part, without distinguishing PINS from PINAS.

Similar to the NBN model, the PIN in a noise scenario should be considered as a sum of different independent components

$$n_{\mathrm{PIN}}(t) = \sum_{i=1}^{N_{\mathrm{PIN}}} c_{\mathrm{PIN},i}(t), \tag{3.93}$$

where every component $c_{\mathrm{PIN},i}$ is wide-sense cyclostationary and can be individually characterized by its parameters such as variance $\sigma^2_{\mathrm{PIN},i}$, arriving time $t_{\mathrm{PIN},i}$, impulse width $\tau_{\mathrm{PIN},i}$ and cyclic period $T_{\mathrm{PIN},i}$ [20, 98], as illustrated in Fig. 3.34.

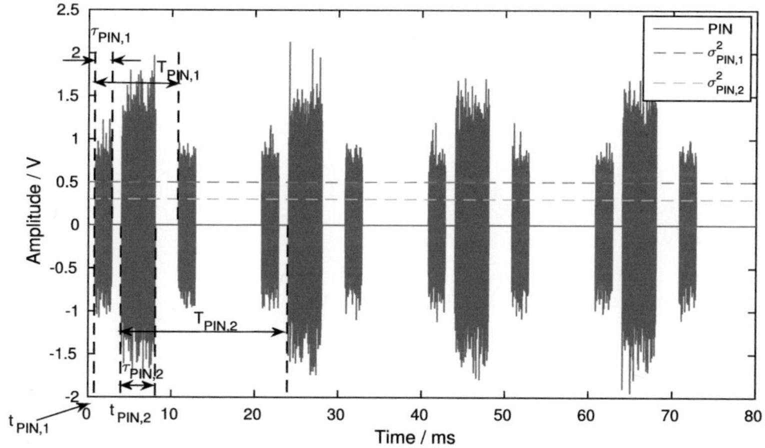

Figure 3.34 Simulated PIN with two independent components $c_{\mathrm{PIN},1}$ and $c_{\mathrm{PIN},2}$, the parameters are labeled

Having been widely used on PIN, the cyclostationary Gaussian model was proposed by *Ohne, Katayama* et al. [44–46, 68], describing any cyclostationary noise at power lines as a colored Gaussian noise with a periodically time-varying variance

$$\begin{aligned}\sigma^2(t,f) &= \sigma^2(t)\alpha(f)\\ &= \frac{\sigma^2(t)ae^{-a|f|}}{2} = \frac{\sigma^2(t+nT_{\mathrm{C}})ae^{-a|f|}}{2}, \quad \forall n \in \mathbb{Z},\end{aligned} \tag{3.94}$$

where $T_{\mathrm{C},i}$ is the cyclic period. The time fluctuation of the variance can be modeled as a periodic amplitude modulation effect, while the frequency dependency $\alpha(f)$, as presented in [44], can be modeled with an LTI filter. Thus, each independent PIN component can be modeled as

$$c_{\mathrm{PIN},i}(t) = \sigma^2_{\mathrm{PIN},i} n_{\mathrm{W,Uni}}(t) * h_{\mathrm{PIN},i}(t), \tag{3.95}$$

where $*$ denotes the convolution operation, i.e.

$$x(t) * y(t) = \int_{-\infty}^{\infty} x(t-\tau)y(\tau)\mathrm{d}\tau, \tag{3.96}$$

and $h_{\mathrm{PIN},i}(t)$ is the impulse response of the LTI filter with the frequency response $H_{\mathrm{PIN},i}(f) = \alpha_i(f)$. This model is also referred to as the temporal cyclostationary model in the form of (3.95) [66], and has been extended by *Nassar* et al. into the so-called spectrotemporal cyclostationary model [65, 66] demoted as

$$c_{\mathrm{PIN},i}(t) = \int_{-\infty}^{\infty} n_{\mathrm{W,Uni}}(t-\tau)h_{\mathrm{PIN},i}(t,\tau)\mathrm{d}\tau, \tag{3.97}$$

where $h_{\mathrm{PIN},i}(t,\tau)$ is an LPTV filter.

As discussed in Sec. 3.1.5.3, an LPTV system can be approximated with an LSTV system. One such approximation was reported by *Lin* et al. in [52], modeling every LTI filter in the LSTV system as an AR filter. Another usual LSTV approximation has been applied in [55, 100], which has an on-off keying structure, described by

$$h_{\mathrm{PIN},i}(t,\tau) = \begin{cases} \sigma^2_{\mathrm{PIN},i}\delta(t-\tau), & t \in [t_{\mathrm{PIN},i} + nT_{\mathrm{PIN},i}, t_{\mathrm{PIN},i} + \tau_{\mathrm{PIN},i} + nT_{\mathrm{PIN},i}] \\ 0, & \text{otherwise} \end{cases}, \quad \forall n \in \mathbb{Z}. \tag{3.98}$$

In the following part of this work, the model (3.98) is taken for PIN.

3.2.5.2 Detecting the Cyclic Frequencies According to the Cyclic Coherence Function

For the extraction, identification and estimation of any independent PIN component $c_{\mathrm{PIN},i}$, the most important parameter is its cyclic period $T_{\mathrm{PIN},i}$, which is usually required as an a-priori knowledge by the signal processing algorithms. Therefore, as the first step of PIN estimation, the cyclic periods of all existing PIN components have to be detected first.

As a powerful tool for cyclostationary signal analysis, the spectral autocoherence was introduced by *Gardner* in [33], which is also known in other literature as the cyclic coherence function (CCF) [1, 101]. For a given signal $x(t)$, its CCF is defined as

$$\begin{aligned} C_{XX}(f,\alpha) &= \frac{S_{XX}(f,\alpha)}{\sqrt{S_{XX}(f-\frac{\alpha}{2})S_{XX}(f+\frac{\alpha}{2})}} \\ &= \frac{\mathrm{E}\{X(f+\frac{\alpha}{2})X^*(f-\frac{\alpha}{2})\}}{\sqrt{\mathrm{E}\{|X(f+\frac{\alpha}{2})|^2\}\mathrm{E}\{|X(f-\frac{\alpha}{2})|^2\}}}, \end{aligned} \tag{3.99}$$

where X is the spectrum of x, X^* is the complex conjugate of X, $S_{XX}(f)$ is the power spectral density (PSD) of x and $S_{XX}(f,\alpha)$ is the so-called cyclic spectral density (CSD) of x. The CSD can also be estimated according to the *Cyclic Wiener-Khinchin Relation* [31] given by

$$S_{XX}(f,\alpha) = \int_{-\infty}^{\infty} R_{xx}^{\alpha}(\tau)e^{-j2\pi f\tau}\mathrm{d}\tau, \tag{3.100}$$

$$R_{xx}^{\alpha}(\tau) = \lim_{T\to\infty}\frac{1}{2T}\int_{-\infty}^{\infty} R_{xx}(t,\tau)e^{-j2\pi\alpha t}\mathrm{d}t, \tag{3.101}$$

where $R_{xx}(t,\tau)$ is the ACF of x. (3.100) is a cyclostationary extension of the well-known *Wiener-Khinchin Theorem* [47].

As a cyclic descriptor, the CCF C_{XX} is limited in the range [0,1], and it is supposed to have a high value at (f_0, α_0), if and only if a CS2 component in x with a cyclic frequency α_0 which has power in its spectrum at the frequency f_0, which can be derived as follows: Let

$$x(t) = x_{\mathrm{c}}(t) + x_{\mathrm{r}}(t), \tag{3.102}$$

where x_{c} is the CS2 part with the cyclic frequencies $\mathbf{A} = \{\alpha_1, \ldots, \alpha_M\}$, and x_{c} is the residual. Then we get

$$\begin{aligned}
&\mathrm{E}\{X(f+\frac{\alpha}{2})X^*(f-\frac{\alpha}{2})\}\\
=&\mathrm{E}\{X_{\mathrm{c}}(f+\frac{\alpha}{2})X_{\mathrm{c}}^*(f-\frac{\alpha}{2})\} + \mathrm{E}\{X_{\mathrm{c}}(f+\frac{\alpha}{2})X_{\mathrm{r}}^*(f-\frac{\alpha}{2})\}\\
&+\mathrm{E}\{X_{\mathrm{r}}(f+\frac{\alpha}{2})X_{\mathrm{c}}^*(f-\frac{\alpha}{2})\} + \mathrm{E}\{X_{\mathrm{r}}(f+\frac{\alpha}{2})X_{\mathrm{r}}^*(f-\frac{\alpha}{2})\}\\
=&\mathrm{E}\{X_{\mathrm{c}}(f+\frac{\alpha}{2})X_{\mathrm{c}}^*(f-\frac{\alpha}{2})\} = \sum_{\alpha_m\in\mathbf{A}} S_{X_{\mathrm{c}}X_{\mathrm{c}}}(f,\alpha)\delta(\alpha-\alpha_m).
\end{aligned} \tag{3.103}$$

To verify this, an artificial noise x is simulated, as shown in Fig. 3.35, which has 4 components n_1 to n_4 of different behavior. n_1 is an aperiodic impulse which lasts 40 µs, n_2 is a periodic impulse series with a cyclic frequency of $\alpha_1 = 1\,\mathrm{kHz}$ and a duty rate of 20%, n_3 is a 3 MHz oscillation, amplitude-modulated by a 100 Hz sinusoidal wave, and n_4 is a WGN.

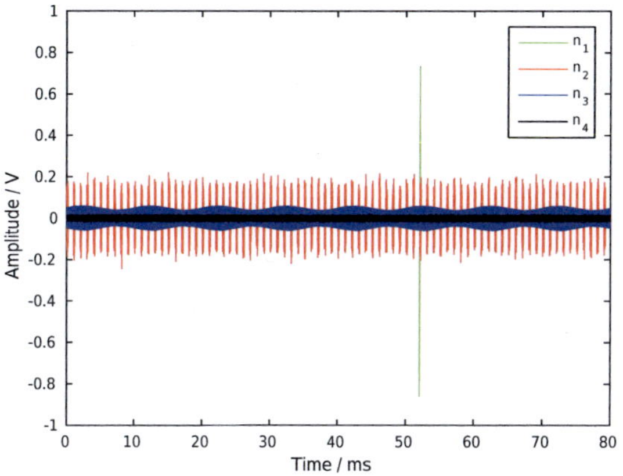

Figure 3.35 Artificial noise with 4 components n_1 to n_4, which were simulated as APIN, PIN, NBN and BGN, respectively

The CCF of this noise is shown in Fig. 3.36. High-valued stripes parallel to the f-axis can be observed as patterns at the cyclic frequencies $\alpha_i = i\alpha_1$, where $i \in \{1, 2, 3, 4, 6\}$, because the rectangular envelope of n_2 includes a set of harmonics of α_1. Additionally, the CCF is also high at $\alpha = 0$, because $C_{XX}(f, 0) \equiv 1$, which can be easily derived from (3.99-3.101). Furthermore, the cyclic frequency 100 Hz of n_3 is insignificant in the CCF.

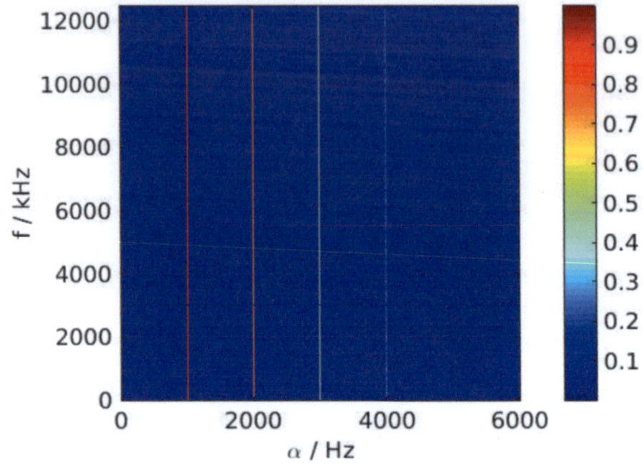

Figure 3.36 The CCF of the noise $x = \sum_{i=1}^{4} n_i$ in Fig. 3.35, the value is significantly high at $\alpha = 0, 1000, 2000, 3000, 4000$Hz and generally low at other cyclic frequencies.

To efficiently detect the striped CCF-pattern of PIN, a cyclic index is designed as the integration of the CCF over f, i.e.

$$\Gamma_X(\alpha) = \int_0^{\infty} C_{XX}(f, \alpha)\mathrm{d}f. \tag{3.104}$$

To verify the sensitivity and selectivity of this descriptor to PIN components and its robustness against noises of other types, four different sample noises are defined as

$$x = n_1 + n_2 + n_3 + n_4; \tag{3.105}$$
$$x_1 = x + 10 \cdot n_1; \tag{3.106}$$
$$x_2 = x + 10 \cdot n_3; \tag{3.107}$$
$$x_3 = x + (10 + 3 \cdot \cos 200\pi t) \cdot n_4, \tag{3.108}$$

where n_1 to n_4 are the noise components in Fig. 3.35, simulated as APIN, PIN, NBN and BGN, respectively. Their cyclic indices are computed and compared, as shown in Fig. 3.37. It can be seen that the cyclic index shows a good sensitivity to PIN, and high robustness against NBN and BGN even when they are modulated by periodic envelopes. But the performance is significantly low when tested with x_1, in which an aperiodic impulse with extremely high amplitude is present. This reveals that the cyclic index is sensitive to high APIN peaks. Therefore, the APIN removal method introduced in Sec. 3.2.4.2 should be first applied, before the cyclic index of the residual is computed, so that the cyclic frequencies can be detected at the peaks.

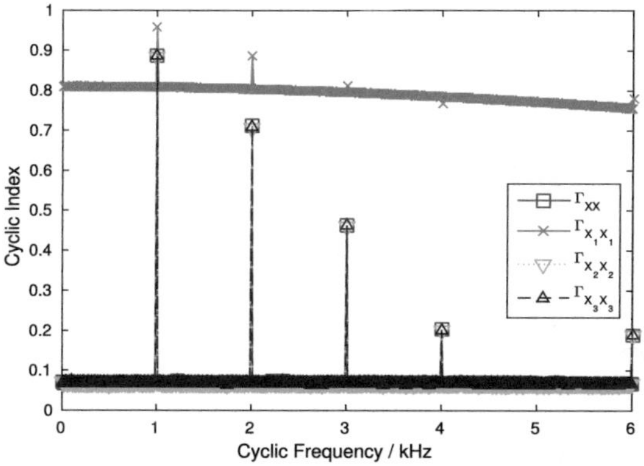

Figure 3.37 Cyclic index of noises x_1 to x_3 from Eqs. (3.105-3.108), $\alpha = 0$ is ignored

3.2.5.3 Extracting Periodic Impulses with the Multiple Cyclic Regression Method

Given the cyclic frequencies of all the CS2 components in a noise measurement, different solutions are available to extract the CS2 part from the noise. *Lin* et al. modeled every CS2 part as the output of an LSTV filter array, estimated every LTI filter in the array with an AR model, and observed a good approximation [52]. A more general CS2 cancellation algorithm for PIN was reported in [101], which makes use of the multiple cyclic regression (MCR) method.

The MCR method was initially proposed by *Bonnardot* et al. in [10], aiming at extracting wide-sense second-order cyclostationary components in mechanical vibration signals. Its efficiency on the PIN components in LV power line noises was then verified in [101] through both real data tests and Monte-Carlo simulations. The principle of this method has its roots in the fact revealed by (3.103), that the CS2 part of a noise has a spectral redundancy, while the rest does not. Thus, a reconstructor can be designed as

$$\hat{X}_{\mathrm{c}}(f) = \sum_{m=1}^{M} X(f-\alpha_m)G_m(f) = X(f)\sum_{m=1}^{M}\delta(f-\alpha_m)G_m(f), \tag{3.109}$$

where M is the amount of cyclic frequencies $\{\alpha_1,\ldots,\alpha_M\}$, each $G_m(f)$ is an LTI filter, or correspondingly represented in the time domain as

$$\hat{x}_{\mathrm{c}}(t) = \sum_{m=1}^{M} e^{j2\pi\alpha_m t}x(t) * g_m(t) = x(t) * \sum_{m=1}^{M} g_m(t)e^{j2\pi\alpha_m t}. \tag{3.110}$$

This operation of frequency-shift filtering (FRESH) was derived by *Gardner* to be equivalent to an LPTV filtering [31, 32] denoted as

$$g(t,\tau) = \sum_{m} g_m(t-\tau)e^{j2\pi\alpha_m\tau}, \tag{3.111}$$

$$g_m(\tau) = \frac{1}{T}\int_{-T/2}^{T/2} g(t,\tau)e^{-j2\pi\alpha_m t}\mathrm{d}t. \tag{3.112}$$

Given all cyclic frequencies $\alpha_1,\ldots,\alpha_M$, the LTI filters G_1 to G_M can be optimized through a recursive least squares (RLS) adaptive algorithm, and an optimal LPTV filter $\hat{G}(f,t)$ is thus obtained. Filtering the measurement x with $\hat{G}(f,t)$, an estimation of the corresponding CS2 component $\hat{x}_{\mathrm{c}}$ can be obtained. Details of the RLS algorithm implementation can be found in [101].

With a given noise measurement $x(t)$, its cyclic frequencies can be detected, according to the cyclic index $\Gamma_X(\alpha)$. Denote all these cyclic frequencies (except for $\alpha = 0$) with a set $\mathbf{A}$, recalling the fact discussed in Sec. 3.2.5.2, that every PIN includes several harmonics in its cyclic spectrum, $\mathbf{A}$ can be divided into several subsets

$$\mathbf{A}_1 = \{\alpha_{1,1}, \alpha_{1,2}, \dots, \alpha_{1,M_1}\}, \tag{3.113}$$
$$\mathbf{A}_2 = \{\alpha_{2,1}, \alpha_{2,2}, \dots, \alpha_{2,M_2}\}, \tag{3.114}$$
$$\dots$$
$$\mathbf{A}_N = \{\alpha_{N,1}, \alpha_{N,2}, \dots, \alpha_{2,M_N}\}, \tag{3.115}$$
$$\mathbf{A} = \bigcup_{n\in\{1,\dots,N\}} \mathbf{A}_n. \tag{3.116}$$

Each subset is a group of harmonics, i.e.

$$\alpha_{n,m} = m\alpha_{n,1} \quad \forall n \in \{1,\dots,N\}, m \in \{1,\dots,M_n\}. \tag{3.117}$$

As an example, simulated noise with two different PIN components is given in Fig. 3.38(a), with its cyclic index analyzed in Fig. 3.38(b).

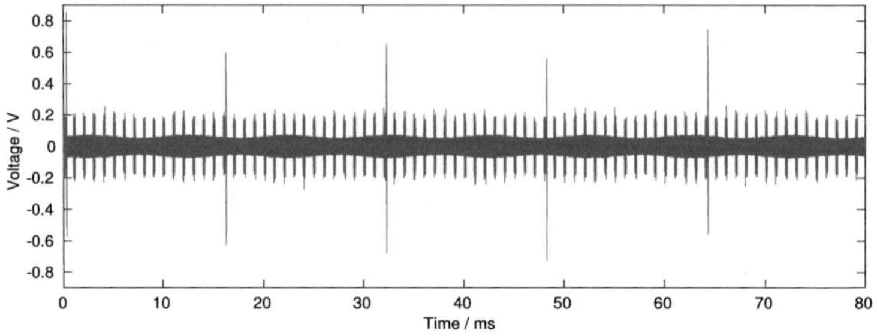

(a) Noise waveform

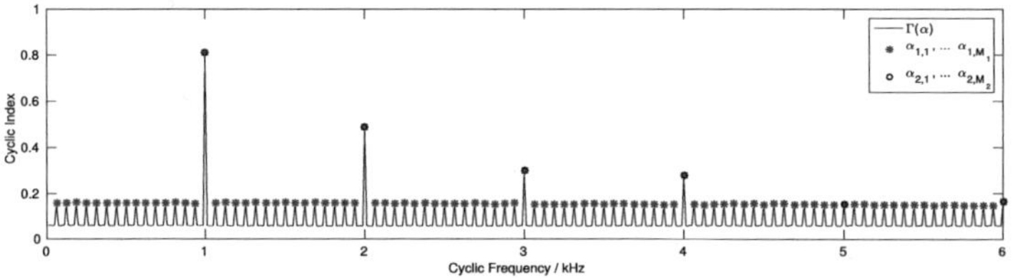

(b) Detecting and grouping cyclic frequencies

Figure 3.38 Analyzing simulated noise with 2 PIN components of different cyclic frequencies

Afterwards, for each $\mathbf{A_n}$, the MCR method is used to optimize the LTI filter array:

$$\hat{\mathbf{G}}_n = [\hat{G}_{n,-M_{\max}}, \hat{G}_{n,-M_{\max}+1}, \ldots, \hat{G}_{n,-1}, \hat{G}_{n,1}, \ldots, \hat{G}_{n,M_{\max}}], \quad (3.118)$$

where each $\hat{G}_{n,m}$ is estimated, according to the cyclic frequency $m\alpha_{n,1}$. $2M_{\max}$ is the dimension of the implemented LTI filter array. Filtering x with the LPTV filter, constructed by this LTI filter array, the corresponding CS2 component estimation $\hat{x}_{cn}$ with main cyclic frequency $\alpha_{n,1}$ can be obtained. Applying the method to the example simulation above, with the detected main cyclic frequencies $\alpha_{1,1} = 62.5\,\text{Hz}, \alpha_{2,1} = 1\,\text{kHz}$, the filter array dimension $M_{\max} = 6$ and implementing every LTI filter as a 10-ordered FIR filter, two CS2 components can be extracted as shown in Fig. 3.39. A good approximation can be observed.

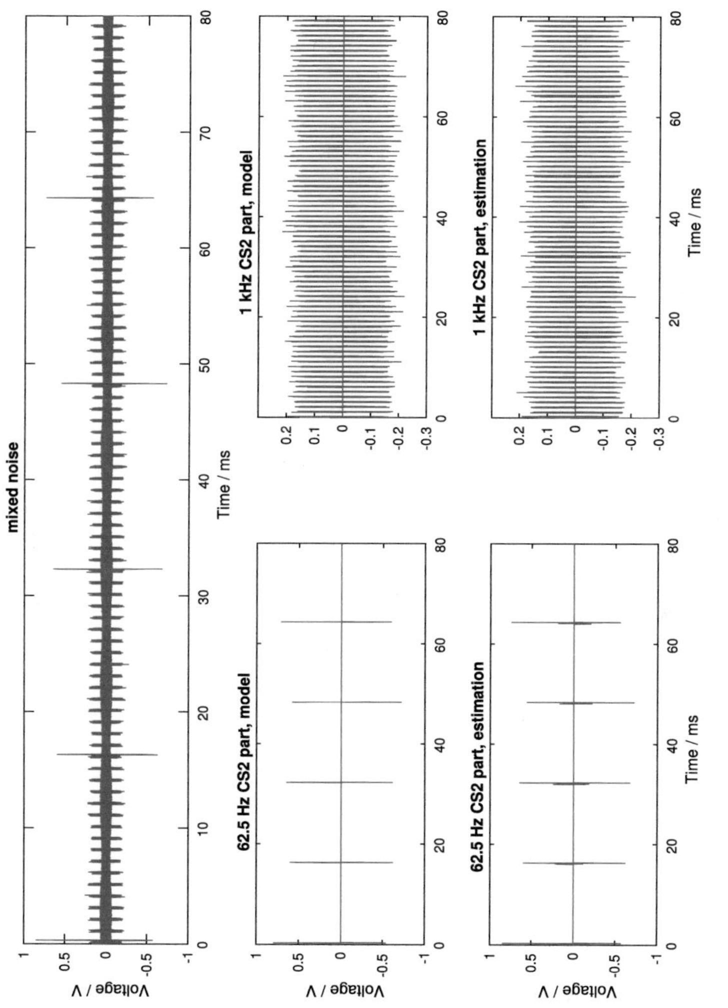

Figure 3.39 Extracting CS2 components from simulated noise

3.2.5.4 Detecting Independent PINs According to the Average Amplitude

After the CS2 component extraction, the independent PINs can be detected. For every CS2 component $\hat{x}_{cn}$ estimated in the last step, the amplitude is averaged through cyclic periods, i.e.

$$\bar{A}_{cn}(t) = \frac{1}{N_{\mathrm{prd}}} \sum_{i=0}^{N-1} |\hat{c}_{cn}(t - \frac{i}{\alpha_{n,1}})| \qquad t \in [0, \frac{1}{\alpha_{n,1}}), \tag{3.119}$$

where $N_{\mathrm{prd}} = \lfloor \alpha_{n,1} t_{\mathrm{meas}} \rfloor$ and t_{meas} is the length of the measurement. Applying a thresholding on the average amplitude, the PIN components with the cyclic frequency $\alpha_{n,1}$ can be detected, as illustrated in Fig. 3.40(b). Based on the detection, the parameters listed in Fig. 3.34 can be conveniently estimated.

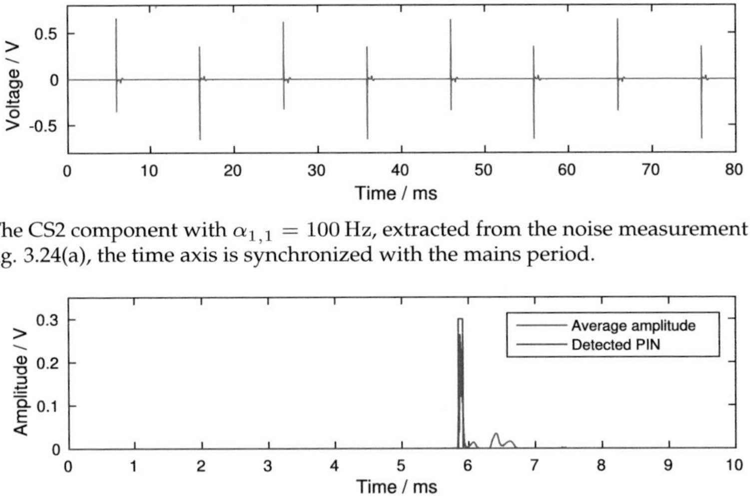

(a) The CS2 component with $\alpha_{1,1} = 100\,\mathrm{Hz}$, extracted from the noise measurement shown in Fig. 3.24(a), the time axis is synchronized with the mains period.

(b) Average amplitude of the extracted CS2 component and the detected PIN

Figure 3.40 Detecting the PIN from an extracted CS2 component

3.2.6 Modeling and Estimating Colored Background Noise

3.2.6.1 Colored Background Noise Model

The colored background noise, as mentioned in Sec. 2.3.3, is a sum of a variety of unknown noise sources with relatively low power. The PSD of CBGN is widely agreed to decay exponentially (in dB) over frequency [16, 21, 40–42, 67], CBGN has been usually considered as time-invariant within up to minutes or even hours [20, 26, 42, 98]. Nevertheless, according to deeper time analysis [16, 21, 45], CBGN proves to be cyclostationary, synchronous with the mains period. Hence, similarly as for PIN, the spectrotemporal cyclostationary model of *Nassar* et al. [65, 66] can also be applied for CBGN in the form

$$n_{\text{CBGN}}(t) = \int_{-\infty}^{\infty} n_{\text{W,Uni}}(t-\tau) h_{\text{CBGN}}(t,\tau) \mathrm{d}\tau, \tag{3.120}$$

where $n_{\text{W,Uni}}$ is a stationary WGN of unity power, and $h_{\text{CBGN}}(t,\tau)$ is an LPTV low-pass filter, synchronous with the mains voltage. Once again, as for the CTF in Sec. 3.1.5.3, an LSTV implementation can be used to approximate $h_{\text{CBGN}}(t,\tau)$, i.e.

$$\begin{aligned} &h_{\text{CBGN}}(t,\tau) \\ &= \begin{cases} h_{\text{CBGN},1}(t), & \tau \in [\tau_1 + nT_{\text{C}}, \tau_2 + nT_{\text{C}}) \\ h_{\text{CBGN},2}(t), & \tau \in [\tau_2 + nT_{\text{C}}, \tau_3 + nT_{\text{C}}) \\ \dots & \\ h_{\text{CBGN},J}(t), & \tau \in [\tau_J + nT_{\text{C}}, \tau_1 + (n+1)T_{\text{C}}) \end{cases} \quad \forall n \in \mathbb{Z}, \end{aligned} \tag{3.121}$$

where J is the size of the LTI filter bank, τ_i is the switching time for the ith LTI filter, and T_{C} is the cyclic period, which is normally one mains period.

3.2.6.2 Estimating the LSTV Filter Model for CBGN

Despite the same representation form, the estimation of the LSTV filter (3.121) for CBGN is more complicated than the estimation of the LSTV filter (3.121) for CTF, because the complex-valued LPTV filter $h_{\mathrm{CBGN}}(t,\tau)$ cannot be directly measured as the time-varying CTF $h(t,\tau)$.

However, generally in practice, the phase information of CBGN is of little interest. This simplifies the task to the estimation of $|H_{\mathrm{CBGN}}(f,t)|$, which is less challenging. According to (3.120), there is

$$S_{\mathrm{CBGN}}(f,t) = S_{\mathrm{W,Uni}}(f,t)|H_{\mathrm{CBGN}}(f,t)|^2, \tag{3.122}$$

where $S_{\mathrm{CBGN}}(f,t)$ and $S_{\mathrm{W,Uni}}(f,t)$ are the instantaneous PSDs of n_{CBGN} and $n_{\mathrm{W,Uni}}$, respectively. After extracting NBN, PIN and APIN from the noise measurement with the aforementioned techniques, $S_{\mathrm{CBGN}}(f,t)$ can be estimated from the residual with time-frequency analyzing methods, such as STFT. As defined before, $n_{\mathrm{W,Uni}}$ is a stationary WGN of unity power, so that $S_{\mathrm{W,Uni}}$ is a constant over frequency and time, i.e.

$$S_{\mathrm{W,Uni}} = \frac{1}{B_{\mathrm{s}}}, \tag{3.123}$$

where B_{s} is the sampling bandwidth of the noise measurement. Therefore

$$|H_{\mathrm{CBGN}}(f,t)| = \sqrt{\frac{1}{S_{\mathrm{CBGN}}(f,t)B_{\mathrm{s}}}}. \tag{3.124}$$

As an example, the CBGN component in the office noise measurement from Figs. 3.22(a) and 3.22(b) is analyzed. An LSTV implementation with 10 LTI filters is estimated, as shown in Fig. 3.41.

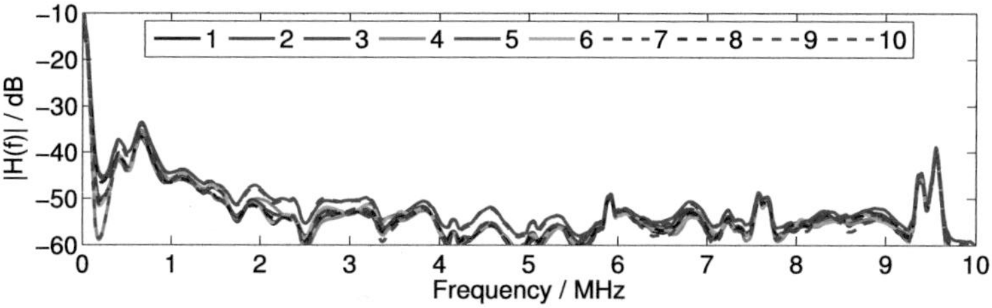

Figure 3.41 The estimated LSTV filter for the CBGN component in the office noise measurement in Fig. 3.22(a), a uniform segmentation with $K = 10$ is applied, so that 10 LTI filters are obtained.

4 Power Line Channel Emulation Concept and its FPGA Implementation

This chapter introduces the concept and implementation of power line channel emulation. First the emulator and the emulator-based PLC testbed are generally studied, explaining the motivation, reviewing the state-of-the-art, and focusing on the challenges and solutions with respect to EMC. Afterwards, as an essential step of cyclostationary channel emulation, the synthesis of a mains phase is discussed. Then, the FPGA-based emulation of CTFs and interference scenarios in PLC networks are introduced sequentially. Sample emulations are given at the end of this chapter for verification.

4.1 High-Speed Power Line Channel Emulator and PLC Testbed

4.1.1 Emulator-Based PLC Testbed: an Overview

As introduced in Chs. 2 and 3, power line channels, in both the aspects of CTF and interference scenario, are frequency selective and time-varying. This leads to a huge challenge for PLC system verification. First of all, the complicated time-varying channel conditions hinder any accurate reproducible testing in a real power grid, and therefore call for an artificial channel reproduction. Next, these channel characteristics are also too difficult to reconstruct accurately with conventional laboratory equipment and analog hardware. Hence, a digital solution is required. Furthermore, although off-line software simulators can be implemented to reconstruct the power line channel behavior flexibly and accurately, their endogenous latencies of interfacing to the devices under test (DUTs) can be critical for performance evaluation. Thus, a real-time hardware solution becomes essential.

An universal solution was proposed by *Götz*: to emulate power line channels with a configurable FPGA-based channel emulator, which was able to emulate usual power line channels flexibly and accurately [36, 37]. Following this idea, different types of power line channel emulators have been developed in the past decade, according to different operation frequency bands of PLC systems. For BB-PLC, *Cañete* et al. proposed an emulator for indoor power line channels in the frequency range up to 30 MHz [15], *Weling* et al. contributed to the multiple-input multiple-output (MIMO) PLC channel emulation [92–94]. For NB-PLC, *Bauer* introduced a channel emulator working in the frequency range 9 - 500 kHz, which was then further improved with a better frequency resolution by *Liu* [57], who also proposed a bidirectional solution and a detailed methodology of PLC system evaluation [55, 58]. However, these existing implementations cannot support a fully parametric emulation of cyclostationary power line channels in the frequency range of 0.15 to 10 MHz, or do not match the characterization approaches discussed in Ch. 3. A new design is therefore motivated.

A typical structure of an uni-directional emulator-based PLC testbed is depicted in Fig. 4.1. The testbed includes an emulator, a server computer, a UPS, two LISNs and two power sockets. The UPS generates a clean mains voltage, which supplies not only the emulator, but also the DUTs connected at the sockets. The emulator is controlled and configured by a server computer, so that it transmits the input signal with a desired CTF and adds desired interferences at the output. The LISNs are used to block direct transmission between the DUTs through the power connection, so that the communication channel is given by the emulator.

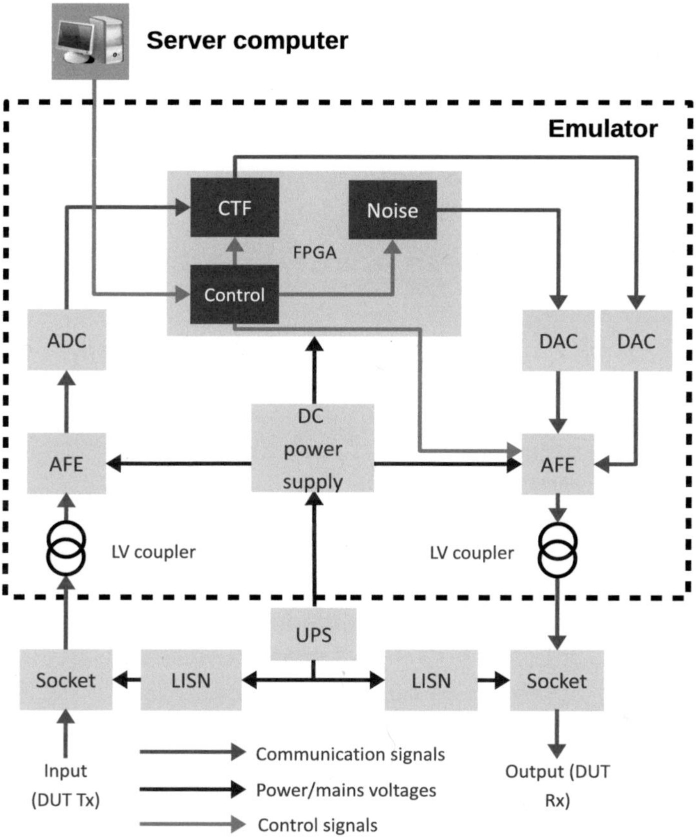

Figure 4.1 Structure of a typical emulator-based PLC testbed

More deeply, the emulator is typically constructed with an FPGA block, a high-speed ADC, two high-speed digital-to-analog converters (DACs), two AFEs, two LV couplers, and a direct current (DC) power supply module. The core component of the emulator is the FPGA block, with which the CTF and noise scenarios are digitally emulated. The control logic is implemented in the FPGA module as well, by which the configuration commands from the server are received, decoded and executed. The AFEs perform general signal conditioning, such as buffering, low-pass filtering and amplification. Particularly, the AFE on the output side also plays the role of mixing the filtered signal and the emulated interferences under a configurable SNR.

4.1.2 EMC Challenges and Solutions

Generally, there are three potential EMC problems in the emulator-based PLC testbed, that may decrease accuracy and reliability:

1. **Crosstalk through unwanted channels**. Besides the channel implemented by the emulator, the signal transmitted by the DUT to the emulator input may also propagate through the air or wire connections to the receiver at the output. This crosstalk brings external parallel transmission paths out of control. This impact is especially significant when a high attenuation is desired and emulated.

2. **Unwanted internal noises**. Almost all analog components with high operational power can generate strong noises, e.g. the power amplifier in the AFE and switching regulators in the DC power supply. Additionally, the FPGA module must generate rectangular clocking signals for its timing, whose frequency can often exceed 100 MHz. These clocks are widely distributed over the entire digital circuitry, and usually are unnegligible with respect to their power. All these noises and clocks can be easily leaking through air and wires, and thus be received by the DUT.

3. **Unwanted external noises**. Environmental noises, such as broadcast radio signals and power line noises leaking from the power cables nearby, may also be received by the DUT.

To avoid influences from these phenomena, the circuit boards must be carefully designed, following strict EMC rules, e.g. the digital and the analog components should be distributed separately and should not share the same ground. Besides, LISNs should be deployed, and the emulator should be built into a proper shell.

4.1.2.1 LISN

As discussed in Sec. 3.2.1.1, an LISN can strongly attenuate high-frequency signals, while passing the mains voltage. Being set between the DUTs, as shown in Fig. 4.1, the two LISNs block the direct signal transmission through the power cable, which could otherwise create strong crosstalk. Moreover, the noise generated by the UPS is also attenuated by the LISNs, so that the external noise at the output of the testbed is reduced.

4.1.2.2 Emulator Shell

Aiming not only at a mechanical protection of the emulator, but also an improvement of its EMC, the shell must be deliberately designed with respect to size, material and wiring.

The wires connecting the DUTs and the testbed, although being normally too short to operate as optimal antennas for frequencies up to 10 MHz, can still transmit/receive a certain power to/from the air. If the input and output ports of the testbed are placed close to each other, a strong coupling between the wires may happen and create significant crosstalk. As the distance rises, the coupling is suppressed. A quantitative analysis is hard here, because we find no far field approximation, but complicated near field radiation, as the dimension of the testbed is much smaller than the wave length (30 to 200 m for 0.15 to 10 MHz). Nevertheless, a reasonably large size of the shell is generally required, and the two AFEs should be placed at different ends of the shell.

The most important property of the material of the shell is the conductivity. A closed shell made of a conductive material, e.g. metal, can shield the noises generated within the emulator well from the DUTs, and hence reduce the impact of unwanted internal noises. However, the potential of the electric field is a constant on the entire surface of such a shell, which reduces the effective distance between the DUTs, and therefore increases the crosstalk. As the transmitting power of most PLC devices is considerably high, the impact of the crosstalk is usually stronger than of the internal noises. Therefore, dielectric materials, such as plastic are preferred for the shell.

Although it is necessary for reasons of safety at electronic devices with high voltage levels, it can be risky for the emulator to be grounded. Analog ground, which is widely distributed within the emulator, can create a strong common mode connection between the input and the output, and cause significant crosstalk, when connected to an extended ground plane. Hence, the entire emulator should not have any protective earth connection.

4.1.3 Emulator Specifications

As the core of a new PLC testbed for narrowband and broadband applications, a LV power line channel emulator is designed with the specifications listed in Tab. 4.1.

Table 4.1 Emulator specifications

Channel	SISO, unidirectional
Emulated frequency range	Up to 10 MHz
Data conversion rate	50 MSPS
Frequency resolution	$<$ 10 kHz
Supported LTI-CTFs per mains period	Up to 10
Time resolution of time varying behavior	1% mains period
SNR adjustment range	$\geq$ 80 dB

4.2 Synthesizing the Mains Phase for Channel Emulation

As far as power line channels are highly synchronous to the mains voltage, it is unavoidable to synthesize the mains phase as a time reference for channel emulation. An ideal mains voltage course is a sinusoidal wave with a constant period, whose phase can be simply synthesized with a digital timer. However, due to unbalance between power consumption and generation, the mains frequency in real power systems always fluctuates slightly (within $\pm 0.05\,\mathrm{Hz}$), but stochastically around its ideal value [23], which raises the complexity of mains phase synthesis.

One solution is to extend the timer with a random interval, which is easy to generate within an FPGA. Two digital implementations of uniformly distributed random numbers (UDRN), based on M-sequences are proposed by *Götz* [36]. *Liu* et al. implemented four different approaches of generating Gaussian random numbers with UDRN into an FPGA, and evaluated them with respect to performance as well as hardware cost [56]. Based on these methods, a random mains frequency with given mean value and variation can be digitally synthesized. This solution is flexible and has no extra hardware requirements except for the FPGA, but is limited in reliability, because not only the power line channel, but also the instantaneous performance of PLC devices is usually synchronous to the mains voltage. Therefore, it can lead to a deviation of system evaluation, if the time reference of the channel emulation is independent of the actual mains voltage, supplied by the UPS.

An alternative is to obtain the actual mains period with a zero-crossing detector, which has been introduced in Sec. 3.1.3. The ZCD is installed to detect the zero-crossings of the mains voltage at the receiving DUT, and its output is sent to the FPGA, so that the length of every mains cycle $T_{\mathrm{m},1}, T_{\mathrm{m},2}, \ldots$ can be recorded. The timer should be synchronized to the detected mains phase once after initialization, and then continuously update its interval, according to

$$\hat{T}_{\mathrm{m},i} = T_{\mathrm{m},i-1} \qquad \forall i \in \mathbb{Z}^+, \tag{4.1}$$

which is used as the length of current synthesized mains cycle. $\hat{T}_{\mathrm{m},i}$ is then quantized into a 14-bit unsigned integer, so that with a timer, a finest resolution of

$$\frac{2\pi}{2^{14}} = \frac{\pi}{8192} \tag{4.2}$$

can be supported for the synthesized mains phase.

4.3 Emulating Power Line Channel Transfer Functions

4.3.1 Implementing Digital Filters: Direct Convolution and Fast Convolution

As discussed in Sec. 3.1.5.3, the transfer function of an LV power line channel can be modeled as an LPTV filter and approximated as an LSTV filter, which is constructed with several LTI filters, and a periodic switching logic. To implement such an LSTV filter in an FPGA, the first step is to implement the LTI filters. Generally, two kinds of digital implementations are available for LTI filters: the direct convolution and the fast convolution. The direct convolution is a conventional solution, which implements the time behavior of the target filter. In case of FIR filters, the output signal is computed as the correlation of the input and the filter coefficients; for IIR filters, loop-back structures are needed. The fast convolution, also known as block convolution or FFT-based convolution [55], processes the input signal block by block, whereby the DFT of each block is calculated and then multiplied with the frequency response of the target filter, so that the filtering is executed in frequency domain. Computing the IDFT of the result, a block of the output signal is obtained. Both solutions are briefly illustrated in Fig. 4.2.

Both implementations have been applied in existing power line channel emulators. For example, the direct convolution was used by *Götz* [36], *Bauer* [6], *Liu* [57] and *Weling* [92], while *Cañete* selected the fast convolution [15]. A performance comparison between the methods was given by *Liu* in [55], pointing out to that the fast convolution can be faster and more efficient than the FIR-based direct convolution if the order of the filter exceeds 40 (in case of I/O in real numbers). According to the system specifications in Table 4.1, using a sampling rate of 50 MSPS, a typical power line CTF with an impulse response of 15 µs requires a filter order over 750; hence, the fast convolution is more efficient. However, an inherent delay within the signal processing chain is also referred to in the same work, as the main disadvantage of the fast convolution method. This latency can introduce a synchronization error for some low-speed PLC systems using the zero-crossings of the mains voltage as synchronizing reference. Nevertheless, aiming at high-speed new PLC technologies, this risk is negligible here. Therefore, the fast convolution is preferred.

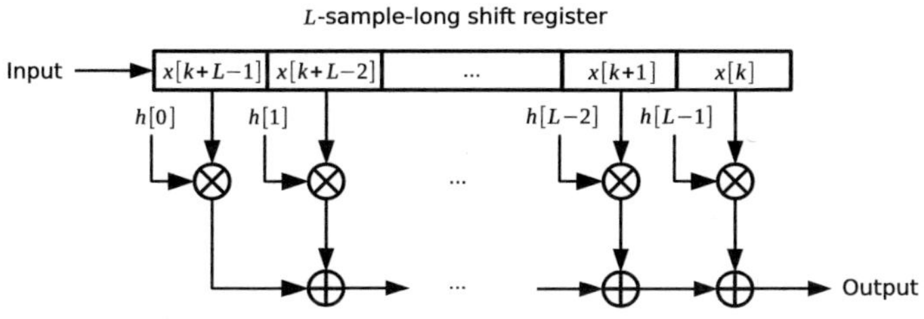

(a) L^{th} order FIR filter implemented in direct convolution

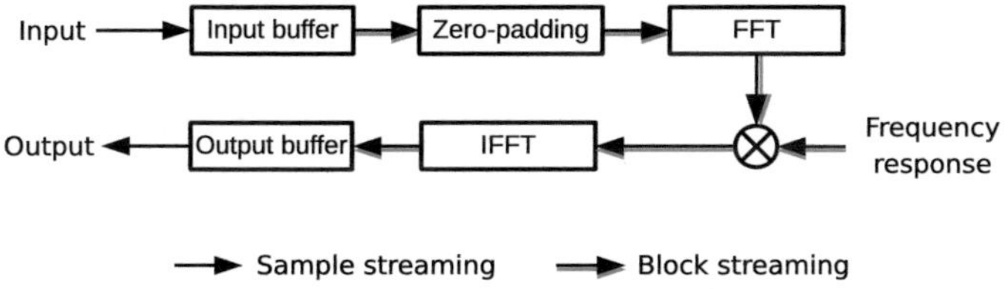

(b) Implementation structure of fast convolution

Figure 4.2 Two possible digital implementations of LTI filters

4.3.2 Implementing Fast Convolution: Overlap-Add and Overlap-Save

As depicted in Fig. 4.2(b), buffers and zero-padding are required for the fast convolution implementation, in order to deal with the timing of segmentation. Consider a case where the input signal is segmented into blocks, each block is L samples long, and the impulse response of the filter is P samples long, the corresponding output of each input block is then $L+P-1$ samples long; hence, its DFT is as well. To obtain such an output DFT block, both the input block and the channel impulse response must be first zero-padded to $L+P-1$ samples, before their DFTs are computed. Additionally, the time interval between each pair of neighboring output blocks is L sampling periods, so that they are always overlapped with $P-1$ samples. According to the method for which the output buffer processes the overlapped part, two different implementations of the fast convolution are available, known as overlap-add (OLA) and overlap-save (OLS) [70].

The difference between OLA and OLS is briefly illustrated in Fig. 4.3. Given two neighboring output blocks y_1 and y_2 which overlap between $K_1 \leq k \leq K_2$, the final output in the overlapped region is given as

$$y_{\mathrm{OLA}}[k] = y_1[k] + y_2[k], \quad \forall k \in [K_1, K_2] \tag{4.3}$$

and

$$y_{\mathrm{OLS}}[k] = y_1[k], \quad \forall k \in [K_1, K_2] \tag{4.4}$$

by OLA and OLS, respectively. A performance comparison between OLA and OLS is reported by *Daher* et al., deriving that they show no difference with respect to errors for time-invariant filters. However, when the filter varies over time, e.g. for LPTV systems, OLA proves to be better for a lower aliasing [18]. Therefore, the overlap-add method is selected in this work.

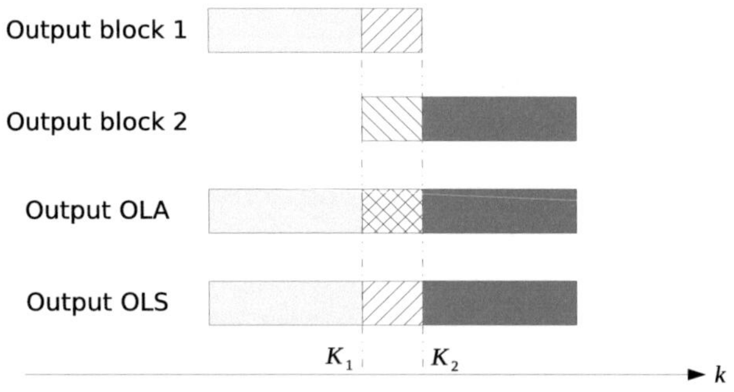

Figure 4.3 Principles of overlap-add and overlap-save

4.3.3 Emulating LPTV Channel Transfer Functions

In order to emulate LPTV CTFs, the fast convolution structure in Fig. 4.2(b) is extended with multiple recorded frequency responses to implement an LSTV filter, as shown in Fig. 4.4. Up to 10 different complex-valued frequency responses are stored in a random-access memory (RAM) bank, each representing one LTI filter. Periodically switching among the stored frequency responses, a cyclostationary CTF can be emulated. To synchronize the switching operation correctly, a table is stored in RAM, recording the filter switching time points with reference to the mains period. According to this table and a synthesized mains phase (introduced in Sec. 4.2), a timing logic switches the filter at desired time instants. As required by the specifications listed in Table 4.1, and supported by the mains synthesizing accuracy, a time resolution of 1% of the mains period is achieved.

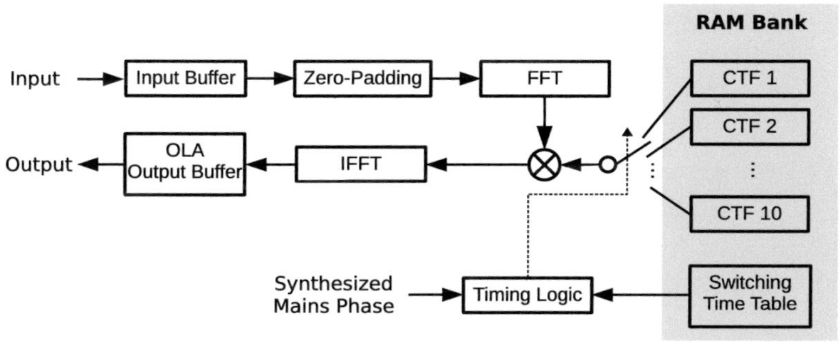

Figure 4.4 An OLA-based LSTV filter to emulate cyclostationary CTFs

4.3.4 Calibrating the CTF Emulation

Besides the fast convolution implemented in the FPGA, other stages in the signal processing chain, such as the AD/DA converters and the AFE, have also impact on the emulated transfer function. This can be described as

$$H_{\text{Emu}}(f,t) = H_{\text{OLA}}(f,t) \cdot H_{\text{Extra}}(f,t), \tag{4.5}$$

where H_{Emu} is the transfer function of configured emulator, H_{OLA} is the transfer function of the fast convolution module, and H_{Extra} is the transfer function of the rest stages aforementioned. Practically, $H_{\text{Extra}}(f,t) = 1$ can not be achieved by optimizing the analog circuits with an affordable effort, so it is never fulfilled that $H_{\text{Emu}} = H_{\text{OLA}}$. Therefore, given a certain target CTF, to correctly configure the fast convolution module, a calibration step is essential.

From (4.5) we have

$$H_{\text{Extra}}(f,t) = \frac{H_{\text{Emu}}(f,t)}{H_{\text{OLA}}(f,t)} = H_{\text{Emu}}(f,t)|_{H_{\text{OLA}}(f,t)=1}. \tag{4.6}$$

Hence, configuring the fast convolution module as a time-invariant zero-phase all-pass filter, $H_{\text{Extra}}(f,t)$ can be measured between the input and the output of the testbed. Then, given a target CTF H_{Target}, the fast convolution should be configured to implement the calibrated CTF

$$H_{\text{OLA}}(f,t) = \frac{H_{\text{Target}}(f,t)}{H_{\text{Extra}}(f,t)}. \tag{4.7}$$

4.4 Emulating the Interference Scenarios

4.4.1 Emulating Narrowband Noise

According to the model (3.59), the NBN should be emulated as the sum of several independent narrowband interferences, whereby each of which has its own envelope (including amplitude), central frequency and bandwidth. To emulate interferences of this kind within an FPGA, a flexible solution based on a phase accumulator and a look-up table (LUT) is available, as illustrated in Fig. 4.5.

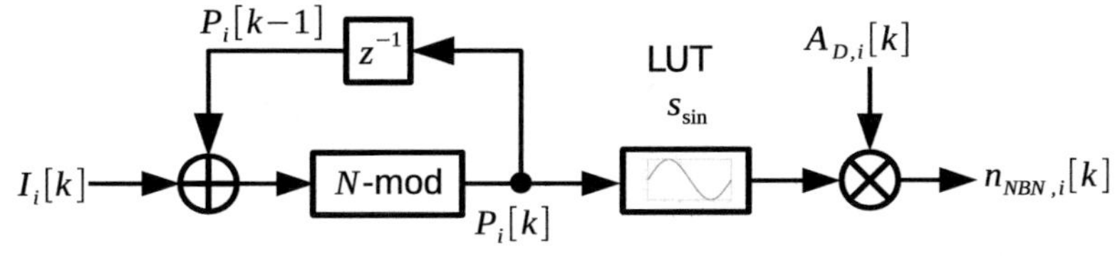

Figure 4.5 Generating narrowband interference with phase accumulation

First, one period of an amplitude-normalized sinusoidal waveform is sampled and stored in an LUT. Denoting the sampled waveform with s_{sin}, there is

$$s_{\text{sin}}[n] = \sin(\frac{2\pi n}{N}), \quad i \in \{0, 1, \dots N-1\}, \tag{4.8}$$

where n and N are the address and the size of the LUT, respectively. To emulate a narrowband interference $n_{\text{NBN},i}$, a phase index $P_i[k]$ is generated by the phase accumulator and updated every clock period, according to the phase increment $I_i[k]$ and the LUT size N, i.e.

$$P_i[k] = \text{Mod}_N(P_i[k-1] + I_i[k]), \quad \forall k \in \mathbb{Z}, \tag{4.9}$$

where k is the time index.

Reading the $P_i[k]^{\text{th}}$ sample of the sinusoidal waveform stored in the LUT, the instantaneous phase of the interference is obtained. This phase is then multiplied with the wanted deterministic time envelope $A_{\text{D},i}$, so that $n_{\text{D},i}$ is synthesized in the form

$$n_{\text{NBN},i}[k] = A_{\text{D},i}[k] \cdot \sin(\frac{2\pi P_i[k]}{N}) = A_{\text{D},i}[k] \cdot s_{\text{sin}}[P_i[k]]. \tag{4.10}$$

Comparing the synthesis formula (4.10) and the model (3.59), the instantaneous frequency $f_i = f_{\text{C},i} + \Delta f_i$ is dependent on the phase increment I_i, the LUT size N and the clock frequency of the FPGA f_{a}, i.e.

$$f_i = \frac{I_i}{N} f_{\text{a}}. \tag{4.11}$$

Thus, a stationary interferer with bandwidth B_i around the central frequency $f_{\mathrm{C},i}$ can be generated by shifting the phase increment in every clock period, according to

$$I_i[k] = I_i[k-1] + D_i[k]\Delta I_i, \tag{4.12}$$

where ΔI_i is the step length of increment shift, and D_i is the shifting direction, with

$$D_i[0] = 1, \tag{4.13}$$

$$D_i[k] = \begin{cases} D_i[k-1] & I_{\mathrm{min},i} \leq I_i[k-1] + D_i[k-1] \leq I_{\mathrm{max},i} \\ -D_i[k-1] & \text{otherwise.} \end{cases} \tag{4.14}$$

So we eventually get

$$f_{\mathrm{C},i} = \frac{I_{\mathrm{min},i} + I_{\mathrm{max},i}}{2N} f_{\mathrm{a}}, \tag{4.15}$$

$$B_i = \frac{I_{\mathrm{max},i} - I_{\mathrm{min},i}}{N} f_{\mathrm{a}}. \tag{4.16}$$

In [36], *Götz* analyzed the performance of this implementation and derived that it supports a frequency resolution, a minimal central frequency and a maximal central frequency of

$$\Delta f = \frac{1}{N} f_{\mathrm{a}}, \tag{4.17}$$

$$f_{\mathrm{C,min}} = \frac{1}{N} f_{\mathrm{a}}, \tag{4.18}$$

and

$$f_{\mathrm{C,max}} = (\frac{1}{2} - \frac{1}{N}) f_{\mathrm{a}}, \tag{4.19}$$

respectively. Also a more efficient design was proposed in the same work, where only a quarter period of the sinusoidal signal is stored in the LUT, so that 75% of the RAM space is saved with a tradeoff of slightly more complexity in the timing logic.

As discussed in 3.2.3.1, the deterministic envelope $A_{\mathrm{D},i}$ is either constant or periodic. Therefore, one simple solution of generating $A_{\mathrm{D},i}$ is to store it in another LUT and to fetch the values cyclically. To improve the efficiency of RAM usage without degrading the precision of quantization, the envelope is first normalized to

$$\tilde{A}_{\mathrm{D},i}[k] = \frac{A_{\mathrm{D,i}[k]}}{\max\limits_{k} |A_{\mathrm{D,i}[k]}|}. \tag{4.20}$$

Thus, one mains period of the normalized envelope $\tilde{A}_{\mathrm{D},i}[k]$ is stored into an LUT, and the peak amplitude $\max\limits_{k} |A_{\mathrm{D,i}[k]}|$ is stored separately as a scaling factor. As shown in Fig. 4.6, with the synthesized mains phase as reference, the normalized envelope value is read from the LUT and scaled, before being sent to the output.

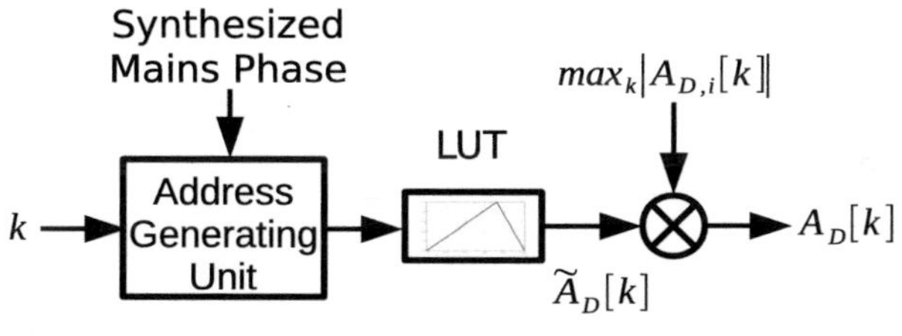

Figure 4.6 Generating time envelope, based on the use of a LUT

To match the specifications in Table 4.1, 10 NBN generators of the structure in Fig. 4.5 are implemented in the emulator, and their outputs are summed, so that up to 10 independent narrowband interferers can be emulated. For each narrowband interferer, a 4000-word, 12-bit RAM is used as LUT to store a sinusoidal wave in 1/4 of its period. Using a 200 MHz sampling rate, a frequency resolution of 12.5 kHz is supported. A 1000-word, 9-bit RAM is used for the envelope LUT, so that a time resolution of 0.1% of the mains period is achieved.

4.4.2 Emulating Aperiodic Impulsive Noise

To emulate APIN according to the occurrence model (3.74), the FPGA architecture shown in Fig. 4.7 is implemented. The amplitude scaling factor σ^2_{APIN} can be obtained from statistical analysis of numerous measurements, and stored in the FPGA as a number. The main effort for such an implementation is generating the WGN $n_{\text{W,Uni}}$ and the occurrence process Φ.

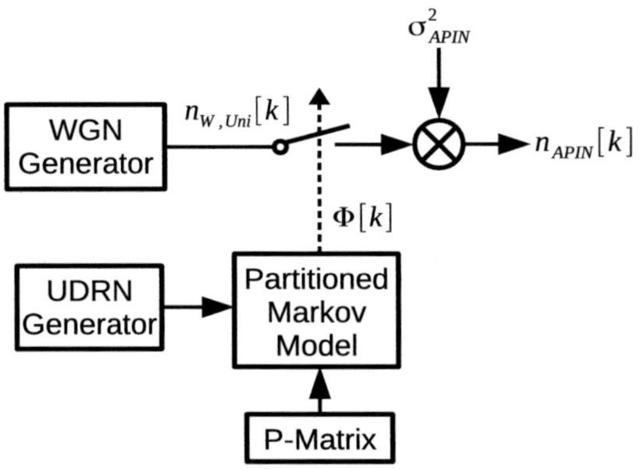

Figure 4.7 Generating APIN in the FPGA

As mentioned in Sec. 4.2, the FPGA implementation of a WGN generator has been deeply discussed in [56]. As one of the four methods evaluated in the mentioned work, the central limit theorem (CLT) method accumulates numbers of independent UDRN to generate WGN. Compared with the Box-Muller method or the polar method and the rejection method, the CLT method is easier to implement and shows satisfying quality of generated noise. Therefore, it is applied here for the generation of WGN.

As discussed in Sec. 3.2.4.1, Φ can be generated by a partitioned Markov chain. An FPGA implementation of this model, proposed by *Götz* in [36], is taken here. Given a transition probability matrix $\mathbf{P}$ describing an APIN scenario with n states (see (3.70) in Sec. 3.2.4.1), a $n \times (n+1)$ cumulative probability distribution matrix is generated in the form

$$P_{\mathrm{C}}(i,j) = \begin{cases} 0 & j = 1 \\ \sum\limits_{k=1}^{j-1} P(i,k) & i \in \{1,2,\ldots,n\}, j \in \{2,3,\ldots,n+1\} \end{cases}. \tag{4.21}$$

The state of the system is initialized as 1, and updated after each clock period according to the generated UDRN $R[k] \in [0,1]$. Residing in state i at time $k-1$, the system switches to state j at time instant k if

$$P_{\mathrm{C}}(i,j) \leq R[k] \leq P_{\mathrm{C}}(i,j+1). \tag{4.22}$$

To accelerate the computation in hardware, the floating probabilities and the UDRN are normalized to 23-bit signed integers in the FPGA implementation. For example, in the cumulative probability distribution matrix

$$\mathbf{P}_{\mathrm{C}} = \begin{bmatrix} -8388608 & 8308104 & 8200104 & 8200104 & 8200104 & 8200104 & 8388607 \\ -8388608 & -8388608 & 8280554 & 8280554 & 8280554 & 8280889 & 8288607 \\ -8388608 & -8388608 & -8388608 & 8300202 & 8300202 & 8300719 & 8388607 \\ -8388608 & -8388608 & -8388608 & -8388608 & 8301050 & 8301405 & 8388607 \\ -8388608 & -8379381 & -8349349 & -8206240 & -8160773 & 8318607 & 8388607 \\ -8388608 & -8279556 & -7924718 & -6237266 & -5699556 & -5699556 & 8388607 \end{bmatrix}, \tag{4.23}$$

six states are presented. The first 4 columns refer to 4 APIN-free states, the columns 5 to 6 refer to 2 APIN occurring states, and the last column refers to probability 1. The implementation of the emulator supports a local storage of up to 3 different $\mathbf{P}_{\mathrm{C}}$ matrices at the same time, so that a convenient switching between different APIN scenarios is possible. As reference APIN scenarios, three different $\mathbf{P}_{\mathrm{C}}$ matrices have been recorded as "strong", "medium" and "weak", respectively, which are listed in Appendices A.1 to A.3.

4.4.3 Emulating Burst Noise

As a special case of APIN, the burst noise is often separately discussed in the emulation of power line noises. A burst event is defined as a package of aperiodic impulses that arrive in a very short time (IAT below 5 ms according to [98]). Such noise is usually modeled as a two-level Markov chain [36, 55], in which each level can be implemented in the same way as for the normal APIN generator. The model and the implementation are illustrated in Fig. 4.8.

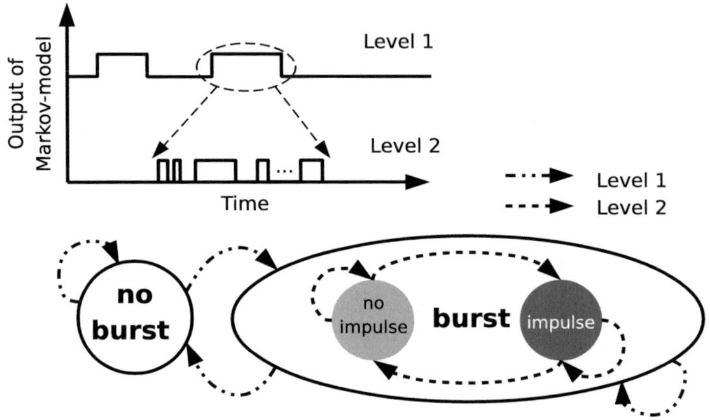

(a) Two-level Markov chain model

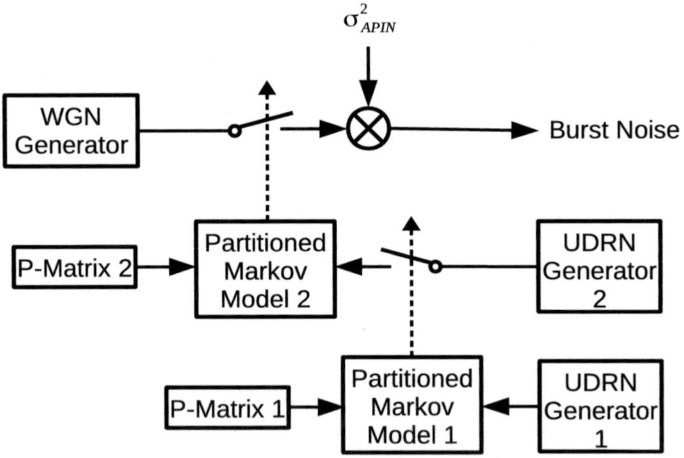

(b) Hardware implementation of the two-level Markov chain model

Figure 4.8 Generating burst noise within the FPGA

4.4.4 Emulating Periodic Impulsive Noise

As (3.93) implies, PIN should be generated as a sum of several independent components. According to the spectrotemporal model (3.97), each component should be generated with an LPTV filter, which normally should be implemented in an LSTV architecture, as shown in Fig. 3.14. However, due to the approximation in (3.98), this LSTV filter can be simplified to an on-off-keying structure, as shown in Fig. 4.9.

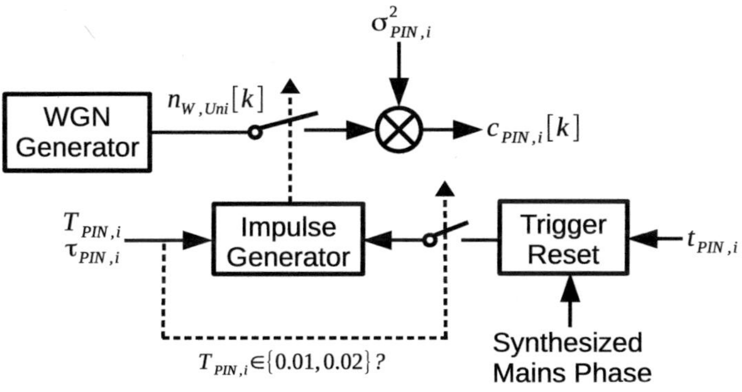

Figure 4.9 Generating a PIN component within the FPGA

According to the IAT $T_{\mathrm{PIN},i}$ and the impulse width $\tau_{\mathrm{PIN},i}$, an impulse generator is responsible to generate a binary periodical impulse series, with which a synthesized WGN is on-off-keying modulated to synthesize the PIN component $c_{\mathrm{PIN},i}$. Particularly, if $T_{\mathrm{PIN},i}$ corresponds to one or one half of the mains period, i.e. $c_{\mathrm{PIN},i}$ is synchronous to the mains, a reset signal is connected to the impulse generator. This reset signal is periodically provided, according to the synthesized mains phase and the impulse arriving time $t_{\mathrm{PIN},i}$, so that the generated impulse series exhibits the desired time offset to the zero-crossings.

In the emulator, 4 such units are implemented to support the generation of up to 4 independent PIN components, and their outputs are accumulated for the final emulation of PIN.

4.4.5 Emulating Colored Background Noise

According to the spectrotemporal cyclostationary model (3.120) and its LSTV approximation (3.121), the CBGN can be emulated with a WGN generator and an LSTV filter. Hence, the FPGA implementation of the LSTV filter discussed in Sec. 4.3.3 is duplicated, receiving a synthesized WGN as its input, so that a cyclostationary CBGN can be obtained at its output. With the same configuration as for the emulation of CTF, up to 10 different states in a mains period can be defined for the CBGN, and a time resolution of 1% of the mains period can be achieved by the switching between states.

4.4.6 Mixing Interferences of Different Types and Calibration

The emulated noises of different kinds are accumulated to construct the final emulation of an interference scenario, as depicted in 4.10. To improve the efficiency of computation, the operation of accumulating is divided into three stages, so that the FPGA's advantage of parallel processing ability is exploited, and thereby the accumulation is accelerated.

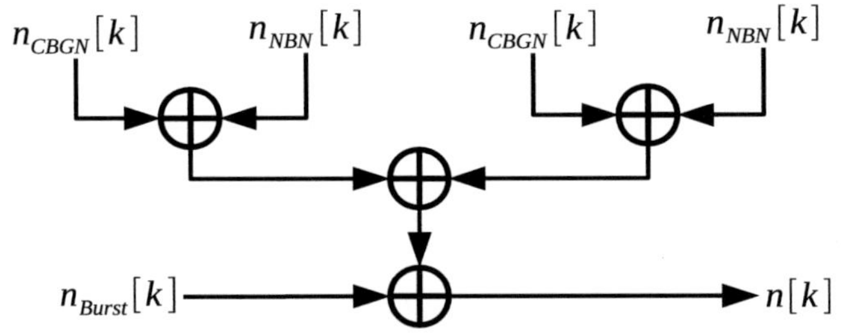

Figure 4.10 Mixing emulated noises of different kinds: Note that the burst noise event is separated from normal APIN, and thus is separately generated here in the emulation.

Similar to the case referred to in Sec. 4.3.4, the emulated interference scenario is also influenced by hardware outside the FPGA, which can be described as

$$N_{\text{Emu}}(f,t) = N_{\text{FPGA}}(f,t) \cdot H_{\text{ExtraN}}(f,t). \tag{4.24}$$

Differing from the case of CTF emulation, neither the AFE at the input of the emulator nor the ADC is included in the signal processing chain of noise emulation. Hence, the hardware have different influences on the CTF emulation and the noise emulation, i.e. $H_{\text{ExtraN}}(f,t) \neq H_{\text{Extra}}(f,t)$, where H_{Extra} is the transfer function of hardware outside the FPGA from (4.5). Thus, an individual calibration is needed for the noise emulation.

Configuring the LSTV filter for CBGN as a time-invariant all-pass filter, and setting all the noise of types other than CBGN to zero, a normalized WGN with unity gain is emulated:

$$N_{\text{FPGA}}(f,t)|_{H_{\text{CBGN}}=1, n_{\text{NBN}}=n_{\text{APIN}}=n_{\text{PIN}}=n_{\text{Burst}}=0} = N_{\text{W,Uni}}(f,t), \tag{4.25}$$

with an expected PSD of

$$S_{\text{FPGA}}(f,t)|_{H_{\text{CBGN}}=1, n_{\text{NBN}}=n_{\text{APIN}}=n_{\text{PIN}}=n_{\text{Burst}}=0} = S_{\text{W,Uni}}(f,t). \tag{4.26}$$

So according to Eqs. (3.123) and (4.24), the PSD measured at the output of the emulator will be

$$S_{\text{Cal}}(f,t) = S_{\text{W,Uni}}(f,t)|H_{\text{ExtraN}}(f,t)|^2 = \frac{|H_{\text{ExtraN}}(f,t)|^2}{B_{\text{S}}}, \tag{4.27}$$

where B_S is the sampling bandwidth of the measurement. Therefore,

$$|H_{\text{ExtraN}}(f,t)| = \sqrt{S_{\text{Cal}} B_{\text{S}}}, \tag{4.28}$$

which should be compensated when configuring the LSTV filter for CBGN and the amplitude factors of other noises.

4.5 Analog Front-End of the Emulator

Two analog front-ends, as mentioned in Sec. 4.1.1, are placed between the AD / DA conversion stages of the emulator and the transmitting /

receiving DUTs, respectively. Due to the differences in functionality, the two AFEs are not symmetrically designed.

4.5.1 AFE at the Input of the Emulator

The AFE at the input of the emulator mainly aims to accomplish the task of anti-aliasing. As depicted in Fig. 4.11, it contains a low-pass filter (LPF) and two buffer amplifiers.

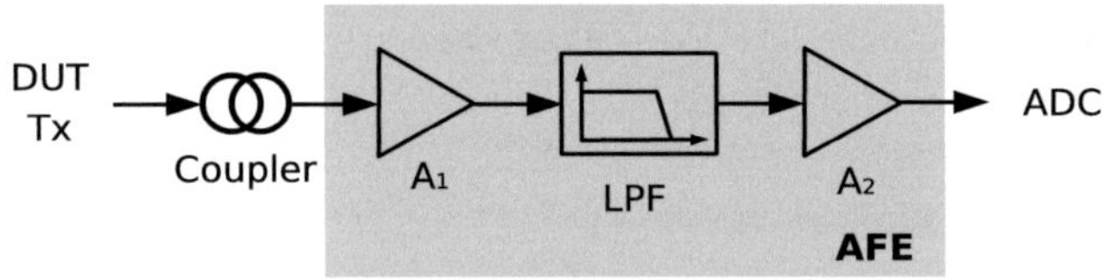

Figure 4.11 Block diagram of the AFE at the input of the emulator, A_1 and A_2 are buffer amplifiers

To simplify the design, the LPF is implemented as a 3rd order Butterworth filter, matched to 50 Ω both at its input and output, as shown in Fig. 4.12 and Tab. 4.2. The magnitude response of the LPF is shown in Fig. 4.13.

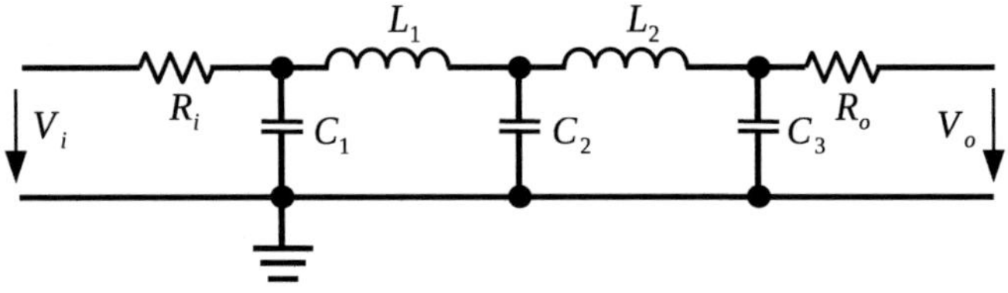

Figure 4.12 Design of the 10 MHz LPF in the AFE

Table 4.2 Parameters of the 10 MHz LPF in the AFE

R_i	C_1	L_1	C_2	L_2	C_3	R_o
50 Ω	220 pF	1.2 µH	680 pF	1.2 µH	220 pF	50 Ω

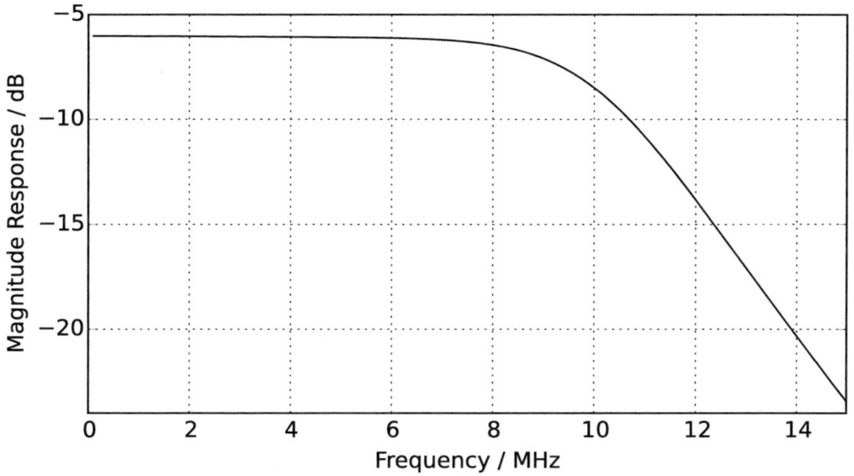

Figure 4.13 Magnitude response of the 10 MHz LPF in the AFE

To avoid coupling losses in magnitude and to compensate the −6 dB attenuation caused by the LPF in its passband, two buffer amplifiers A_1 and A_2 are deployed, both configured to have high input impedance and low output impedance. Their gains are set to

$$|G_1| = |G_2| = 3\,\mathrm{dB} \tag{4.29}$$

4.5.2 AFE at the Output of the Emulator

Compared with the AFE at the input of the emulator, the AFE at is more complex in functionality and structure. First of all, it has two channels, one for the communication signal, filtered by the emulated power line channel, and the other one for the emulated noise. In each channel, a reconstruction filter is needed. Second, both channel outputs must be mixed in a configurable ratio to obtain the complete power line channel output. Furthermore, the mixed signal requires an amplification of power before it is sent to the receiving DUT over a LV PLC coupler. The block diagram is given in Fig. 4.14.

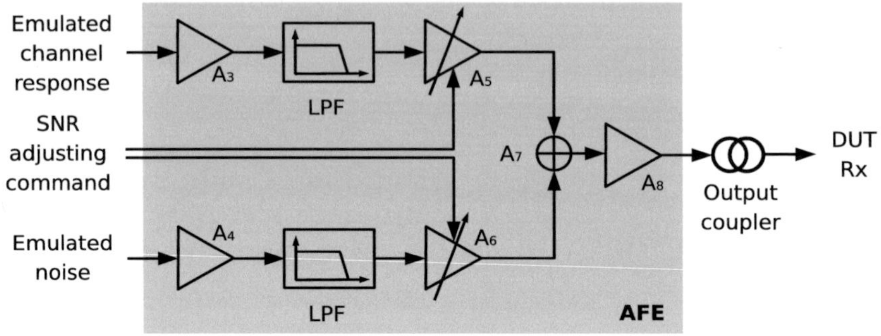

Figure 4.14 Block diagram of the AFE at the output of the emulator, A_3 and A_4 are buffer amplifiers, A_5 and A_6 are variable gain amplifiers (VGAs), A_7 is a summing amplifier and A_8 is a power amplifier.

The LPFs are implemented with the same design as in Fig. 4.12 and Tab. 4.2. Before each LPF, a buffer amplifier is deployed to maximize the input impedance and to compensate the attenuation in the passband. The gains of the buffer amplifiers are set to

$$G_3 = G_4 = 6\,\mathrm{dB}. \tag{4.30}$$

When testing a PLC system under a certain CTF and a certain interference scenario, the dependency between the system performance and the SNR at the receiver is generally of interest. A flexible adjustment of SNR is therefore required. Although this process can be digitally executed in the FPGA, it leads to an increase of quantization errors under limited data word length. Digitally synthesizing a signal s with the peak amplitude m_p, it can be quantized with a N-bit signed integer, i.e. between $L = 2^N$ quantization levels. Assuming that the quantization error q is zero-mean uniformly distributed, the signal-to-quantization-noise ratio is

$$\mathrm{SQNR} = \frac{\mathrm{E}\{s^2\}}{\mathrm{E}\{q^2\}} = \frac{\mathrm{E}\{s^2\}}{\int_{-m_\mathrm{p}/L}^{m_\mathrm{p}/L} q^2 \frac{L}{2m_\mathrm{p}} \mathrm{d}q} = \frac{3P_s}{m_\mathrm{p}^2} \times 2^{2N}, \tag{4.31}$$

where $P_s = 3\mathrm{E}\{s^2\}$ is the signal power. When digitally performing SNR adjustment in the FPGA, N and therefore m_p are fixed by the input/output range of the DAC. Hence, only a decrease of P_s is possible, which leads to a sinking SQNR, according to (4.31).

Therefore, the SNR adjustment is implemented on the AFE at its output with a VGA after each LPF (A_5 and A_6 in Fig. 4.14). With digital commands generated by the FPGA, both VGAs can be separately turned on/off. Their gains G_5 and G_6 can also be configured independently of each other, in the range –40 to 0 dB. Thus, the SNR adjustment range of 80 dB required in Tab. 4.1 is achieved. The separately attenuated channel outputs are then added by a summing amplifier, and sent to a power amplifier, which supplies sufficient power to drive the final output.

4.6 Verification

4.6.1 Verification of the CTF Emulation

4.6.1.1 Verification of the LTI CTF Emulation

As the basics of CTF emulation, the emulation of LTI systems must be first verified. As the first step, an LIT all-pass system was emulated with the testbed. Fig. 4.15. shows the emulation result before and after the calibration, which has been discussed in Sec. 4.3.4. The error between the emulated system and an ideal all-pass system is limited to 2.5 dB and 1.5 dB before and after the calibration, respectively.

Afterwards, a reference LTI power line channel was synthesized with the Zimmermann's model (3.36) and emulated with the calibrated testbed. The result is shown in Fig. 4.16. The emulated CTF matches the model well up to 4 MHz, while the attenuation of the narrow notch around 4.5 MHz is not efficiently achieved, and the attenuation in the second pass-band around 6 MHz is also higher than expected. This error can be explained by the windowing effect, which can be improved by increasing the length of the FFT implementation and selecting a better window shape.

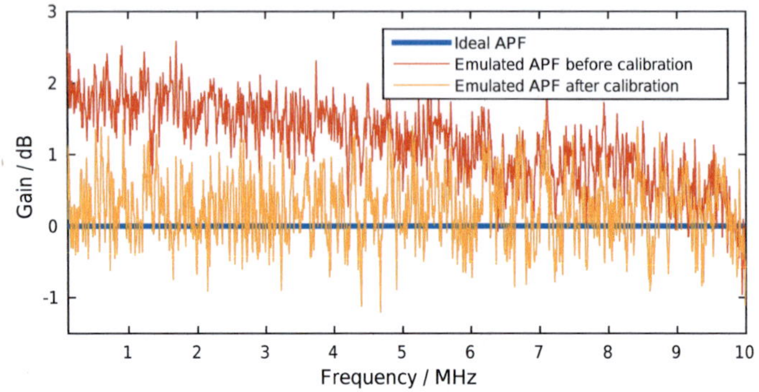

Figure 4.15 Results of emulating an all-pass system with the emulator-based PLC testbed

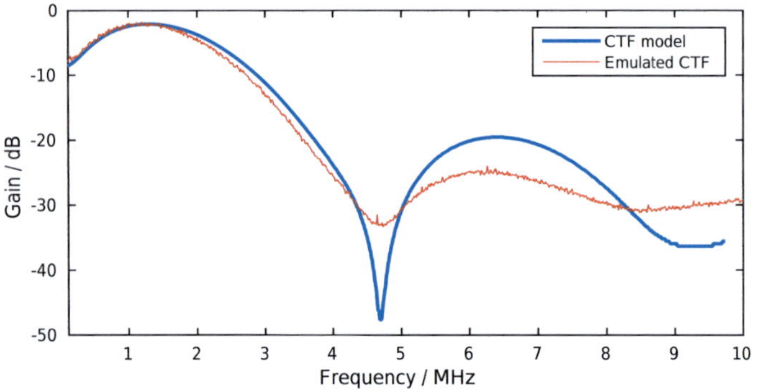

Figure 4.16 Results of emulating a synthesized CTF with the emulator-based PLC testbed

4.6.1.2 Verification of LSTV CTF Emulation

To verify the emulation of LSTV CTFs, four different reference LTI CTFs were first synthesized with the model according to (3.36), as shown in Fig. 4.17. A LSTV CTF which periodically switched among the three reference CTFs was emulated in the emulator, the switching time is given in Tab. 4.3. As shown in Fig. 4.18, a time variation was successfully realized as desired.

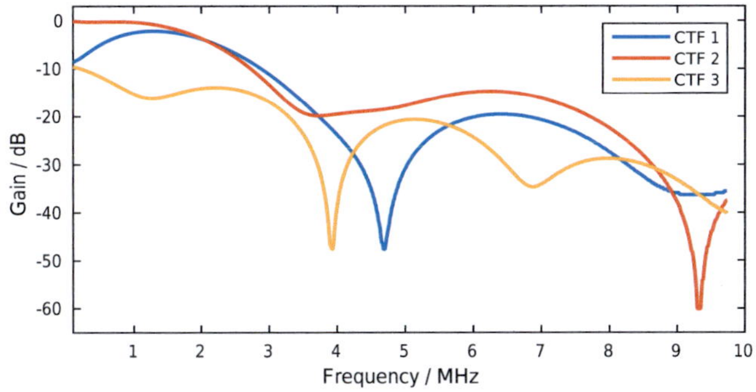

Figure 4.17 Three different reference LTI channel transfer functions

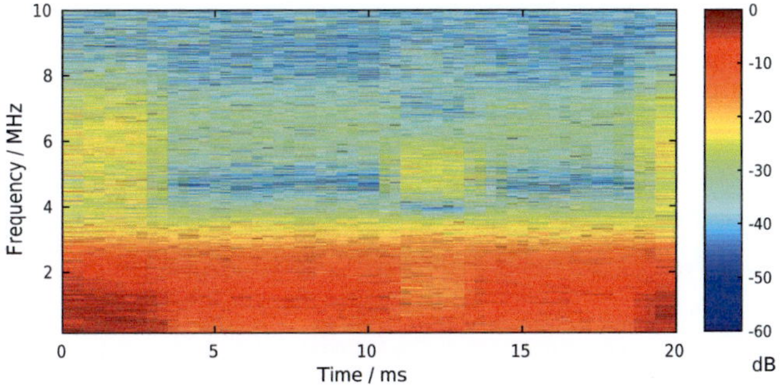

Figure 4.18 Emulated LSTV channel transfer function.

Table 4.3 Emulator configuration: duration of each CTF status in a cyclic period with respect to the mains period T_m.

Status	CTF 1	CTF 2	CTF 1	CTF 3
Duration	$0.25T_\mathrm{m}$	$0.25T_\mathrm{m}$	$0.35T_\mathrm{m}$	$0.15T_\mathrm{m}$

4.6.2 Verification of Inteference Scenario Emulation

4.6.2.1 Verification of NBN Emulation

To verify NBN emulation, narrowband interference with a central frequency of 6 MHz, a bandwidth of 40 kHz, a power of –27 dBV^2, and a triangle envelope of 20 ms period was emulated. The result was measured in the time domain as well as in the frequency domain, as shown in Fig. 4.19.

4.6.2.2 Verification of APIN and Burst Noise Emulation

Aperiodical impulsive noise was emulated, according to the cumulative probability matrix (A.1), resulting in the waveform shown in Fig. 4.20.

A burst noise was emulated, according to the cumulative probability matrices (A.1) and (A.3) for the first and second levels of Markov processes, respectively. The result is shown in Fig. 4.21.

4.6.2.3 Verification of the PIN Emulation

To verify the emulation of PINS and PINAS, three different PINs were designed, as listed in Tab. 4.4, and emulated in two scenarios. The first scenario includes $c_{\mathrm{PIN},1}$ and $c_{\mathrm{PIN},2}$, which are both synchronous to the mains period; in the second scenario only a PINAS $c_{\mathrm{PIN},3}$ is generated. Analyzing the emulation results shown in Fig. 4.22, the parameters of the emulated PINs vary from the desired ones within 3%.

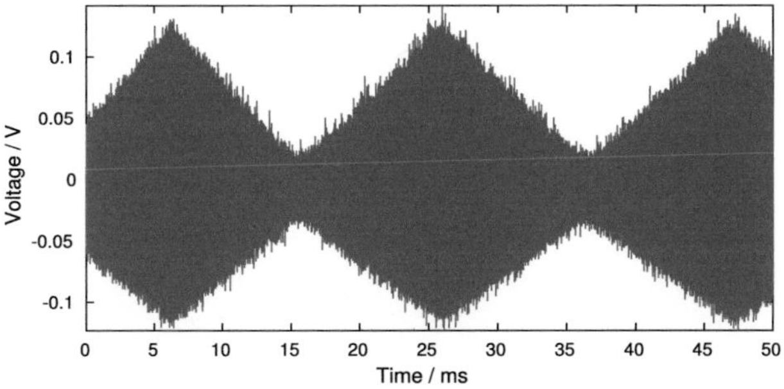

(a) Waveform of the emulated NBN: A triangle envelope is successfully realized.

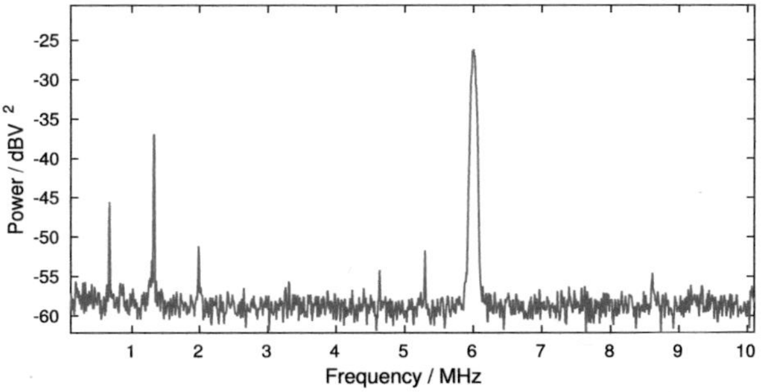

(b) Spectrum of the emulated NBN: Note that the three narrow peaks between 0.5 to 2 MHz were generated by the measuring device. The measured 3-dB bandwidth of the peak around 6 MHz is between 35 to 45 kHz

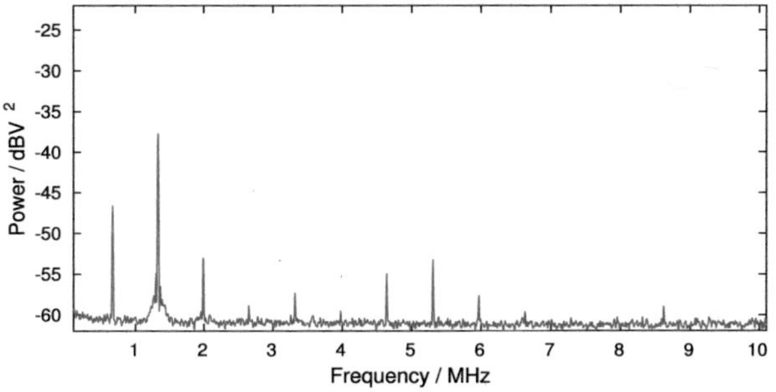

(c) Spectrum of the noise generated by the measuring device

Figure 4.19 Verification of NBN emulation

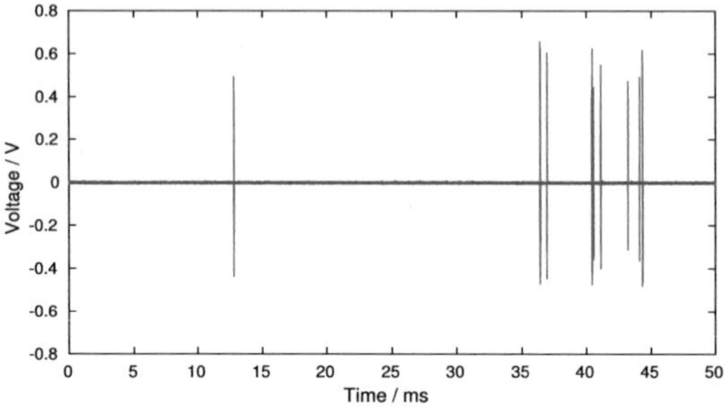

Figure 4.20 APIN emulated under the of "weak" presence configuration

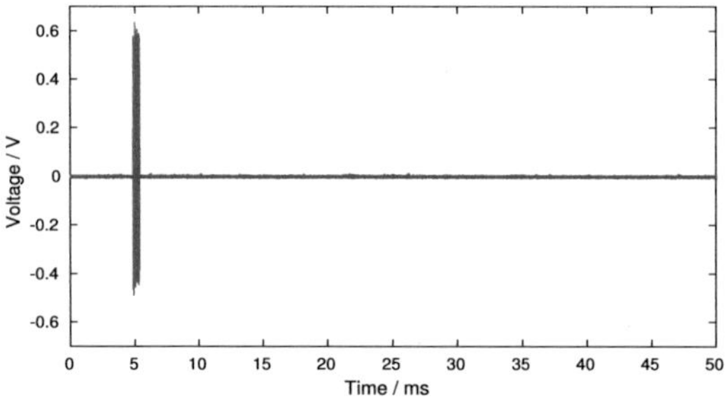

(a) A segment of a waveform with a burst

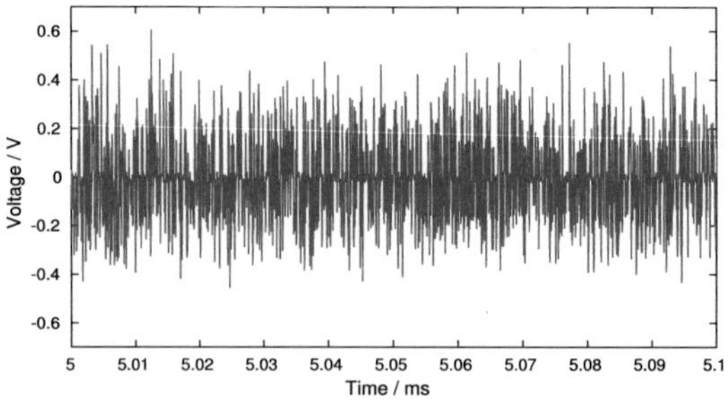

(b) A part of the burst

Figure 4.21 Burst noise emulated under the configuration of "weak" presence of bursts and "strong" presence of impulses in every burst

Table 4.4 Parameters of three different PINs $c_{\mathrm{PIN},1}$ to $c_{\mathrm{PIN},3}$: The first two are synchronous to the mains phase, while the last one is asynchronous.

i	$T_{\mathrm{PIN},i}$	$\tau_{\mathrm{PIN},i}$	$t_{\mathrm{PIN},i}$	$\sigma^2_{\mathrm{PIN},i}$
1	$0.5T_{\mathrm{m}}$	20 ms	$0.05T_{\mathrm{m}}$	–60 dBV^2
2	$0.5T_{\mathrm{m}}$	100 ms	$0.15T_{\mathrm{m}}$	–30 dBV^2
3	340 µs	20 µs	-	–25 dBV^2

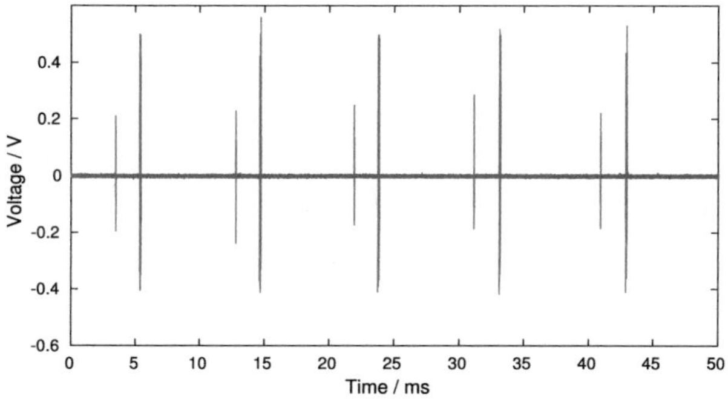

(a) Two emulated PINs, synchronous to the mains voltage: Note that the mains period exhibits a slight fluctuation in this measurement.

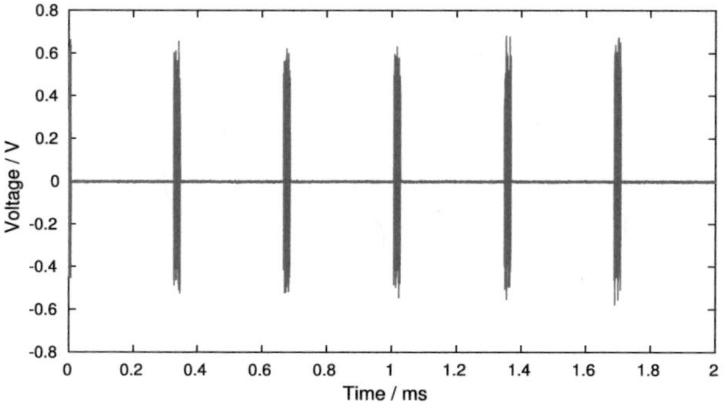

(b) An emulated PINAS

Figure 4.22 Emulated periodical impulsive noises

4.6.2.4 Verification of CBGN Emulation

To verify the emulation of LSTV CBGN, CBGN was generated by filtering WGN with an LSTV filter, which cyclically switched its status among the three transfer functions in Fig. 4.17. The duration of each status in every mains period is listed in Tab. 4.5. As depicted in Fig. 4.23, the result matches the configuration well.

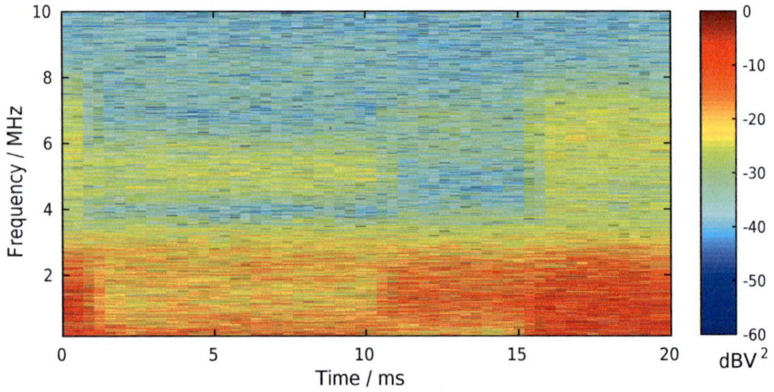

Figure 4.23 Emulated LSTV channel transfer function

Table 4.5 Emulator configuration: duration of each CBGN status in a cyclic period with respect to the mains period T_{m}

Status	CTF 1	CTF 2	CTF 3
Duration	$0.25T_{\mathrm{m}}$	$0.25T_{\mathrm{m}}$	$0.5T_{\mathrm{m}}$

4.6.3 Verification of the SNR Adjustment

To verify the function of SNR adjustment, the emulator was configured with a time-invariant CTF with low-pass behavior and a simple noise scenario with a reference CBGN and a single-frequency oscillation at 6 MHz as NBN. No impulsive noise was emulated. WGN was generated with a laboratory waveform generator to stimulate the emulated channel.

The output of the testbed was then measured 4 times under different configurations of SNR adjustment, which are listed in Tab. 4.6. The result depicted in Fig. shows that the emulated noise and channel response can be successfully mixed under different SNR values.

Table 4.6 Emulator configurations: gains of the amplifiers A_5 and A_6 in Fig. 4.14

Configuration	Scene 1	Scene 2	Noise	Transmission
G_5	0 dB	0 dB	0[a]	0 dB
G_6	0 dB	–10 dB	0 dB	–30 dB

[a] A_5 was completely turned off under this configuration

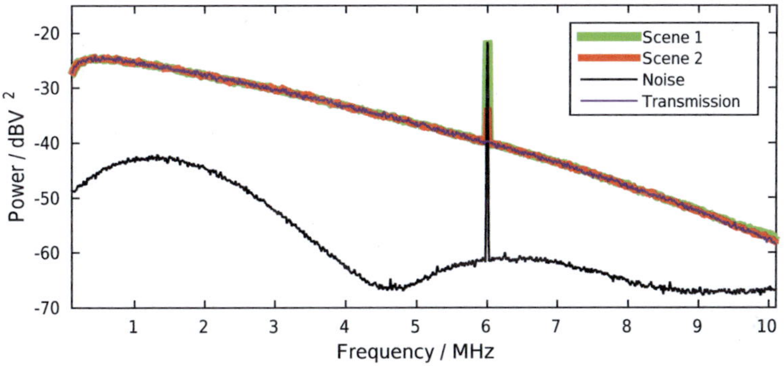

Figure 4.24 PSD measured at the output of the testbed under the configurations listed in Tab. 4.6

5 Conclusions

In this thesis, approaches have been developed to characterize the CTFs and interference scenarios of LV power line networks in the frequency range 0.15 to 10 MHz, including measuring methods and parameter estimation algorithms. Meanwhiles, a PLC testbed, based on a high-speed power line channel emulator, has been developed to support a flexible PHY evaluation of PLC devices in a laboratory environment. Efforts have been made mainly with the following aspects:

- Measuring cyclostationary CTFs
- Modeling cyclostationary CTFs
- Noise decomposition and parameter estimation
- Emulating cyclostationary channels

5.1 Measuring Cyclostationary CTFs

A distributed CTF measuring system was taken to reduce the difficulty of setup and calibration. As candidates for sounding signals, PN sequences and dual sweeping tones were compared under the same constrains of channel conditions, sounding power, and overall measuring duration. Simulations showed that the accuracy and robustness of the measuring system varies only very slightly when using the two different sounding signals. A hybrid synchronizing framework, combining the zero-crossing detector and the GPS reiceivers was designed to synchronize the distributed CTF measuring system.

5.2 Modeling Cyclostationary CTFs

The classic LTI model of power line CTFs was extended step-by-step into a LPTV version, and then a LSTV approximation. Two segmentation methods were designed to transfer a LPTV CTF model into its LSTV approximation: the uniform segmentation, which is simple to implement, and the clustering-based method, which is better in flexibility and accuracy.

5.3 Noise Decomposition and Parameter Estimation

According to the time-frequency characteristics of different noise types, a set of robust methods has been developed to decompose raw measurements of power line noise into separate interferences of different types. A modified version of Welch's method was applied to detect the central frequencies of existing NBNs, so that a time-frequency domain method by *Liu* can follow to extract them and estimate their envelopes. Based on the EMD, a time domain algorithm was developed to detect and extract APIN pulses. A cyclic index was defined, based on the cyclic coherence function, in order to realize a robust detection of cyclic frequencies in time series. The MCR method was then used to extract every PIN component. Making use of the methods designed for the CTF characterization, the modeling of CBGN as LSTV processes was also achieved.

5.4 Emulating Cyclostationary Channels

A LSTV filter architecture was implemented into an FPGA to realize the emulation of cyclostationary CTFs and CBGNs. With help of a zero-crossing detector, the phase of the mains voltage was successfully obtained, so that the CTF, the CBGN and the PINNS could be emulated in a correct synchronization. Besides, analog circuits were developed to accomplish the tasks of coupling, general signal conditioning and SNR adjustment.

5.5 Future Work

Due to the limited time, the designed measuring system has not been implemented in hardware yet. Hence, no measurement data have been obtained. It may be an interesting subject to build a large database of channel measurements with the proposed measuring methods, and to define reference channels in the concerned frequency range. Besides, the access impedance plays an important role in the time variation of the CTF, and can be worth separate investigations in further work.

Appendix

A Reference APIN Matrices

Here follow three cumulative probability matrices, which represent the reference APIN scenarios of "weak", "medium" and "strong" presences, respectively.

A.1 Weak

$$\begin{bmatrix} -8388608 & 8308104 & 8200104 & 8200104 & 8200104 & 8200104 & 8388607 \\ -8388608 & -8388608 & 8280554 & 8280554 & 8280554 & 8280889 & 8288607 \\ -8388608 & -8388608 & -8388608 & 8300202 & 8300202 & 8300719 & 8388607 \\ -8388608 & -8388608 & -8388608 & -8388608 & 8301050 & 8301405 & 8388607 \\ -8388608 & -8379381 & -8349349 & -8206240 & -8160773 & 8318607 & 8388607 \\ -8388608 & -8279556 & -7924718 & -6237266 & -5699556 & -5699556 & 8388607 \end{bmatrix} \quad (A.1)$$

A.2 Medium

$$\begin{bmatrix} -8388608 & 1388104 & 1388104 & 1388104 & 5388104 & 7388104 & 8388607 \\ -8388608 & -8388608 & 2387097 & 3387097 & 4387097 & 7387097 & 8288607 \\ -8388608 & -8388608 & -8388608 & 4982766 & 5982766 & 7984947 & 8388607 \\ -8388608 & -8388608 & -8388608 & -8388608 & 5017998 & 7020179 & 8388607 \\ -8388608 & -8388105 & -8387434 & -8376864 & -8310762 & 7388607 & 8388607 \\ -8388608 & -8377703 & -8362939 & -8130742 & -6684043 & -6684043 & 8388607 \end{bmatrix} \quad (A.2)$$

A.3 Strong

$$\begin{bmatrix} -8388608 & -1677722 & -1677722 & -1677722 & -1677722 & 4194303 & 8388607 \\ -8388608 & -8388608 & -1677722 & -1677722 & -1677722 & 4194303 & 8288607 \\ -8388608 & -8388608 & -8388608 & -1677722 & -1677722 & 4194303 & 8288607 \\ -8388608 & -8388608 & -8388608 & -8388608 & -1677722 & 4194303 & 8288607 \\ -8388608 & -8379386 & -8349349 & -8206240 & -8160773 & 8388607 & 8388607 \\ -8388608 & -8279556 & -7924718 & -6237266 & -5699556 & -5699556 & 8388607 \end{bmatrix} \quad (A.3)$$

Bibliography

[1] **Antoni, J.** *Cyclic Spectral Analysis in Practice*. In: *Mechanical Systems and Signal Processing* 21.2 (2007), pp. 597–630.

[2] **Banwell, T. and Galli, S.** *On the Symmetry of the Power Line Channel*. In: *IEEE International Symposium on Power Line Communications and Its Applications (ISPLC)*. IEEE, 2001, pp. 325–330.

[3] **Banwell, T. and Galli, S.** *A Novel Approach to the Modeling of the Indoor Power Line Channel Part I: Circuit Analysis and Companion Model*. In: *IEEE Transactions on Power Delivery* 20.2 (2005), pp. 655–663.

[4] **Barmada, S., Musolino, A., and Raugi, M.** *Innovative Model for Time Varying Power Line Communication Channel Response Evaluation*. In: *IEEE Journal on Selected Areas in Communications* 24.7 (2006), pp. 1317–1326.

[5] **Barnes, J. S.** *A Physical Multi-Path Model for Power Distribution Network Propagation*. In: *Proc. 1998 International Symposium on Power Line Communications and Its Aplications*. 1998.

[6] **Bauer, M., Liu, W., and Dostert, K.** *Channel Emulation of Low-Speed PLC Transmission Channels*. In: *IEEE International Symposium on Power Line Communications and Its Applications*. IEEE. 2009, pp. 267–272.

[7] **Beichelt, F.** *Stochastische Prozesse für Ingenieure*. Springer, 1997.

[8] **Berger, L. T. et al.** *Introduction to Power Line Communication Channel and Noise Characterisation*. In: *MIMO Power Line Communications: Narrow and Broadband Standards, EMC, and Advanced Processing*. CRC Press, 2014, pp. 3–37.

[9] **Berger, L. T., Schwager, A., and Escudero-Garzás, J. J.** *Power Line Communications for Smart Grid Applications*. In: *Journal of Electrical and Computer Engineering* (2013), p. 3.

[10] **Bonnardot, F., Randall, R., and Guillet, F.** *Extraction of Second-Order Cyclostationary Sources — Application to Vibration Analysis.* In: *Mechanical Systems and Signal Processing* 19.6 (2005), pp. 1230–1244.

[11] **Chaouche, H. et al.** *Time Domain Modeling of Powerline Impulsive Noise at Its Source.* In: *Journal of Electromagnetic Analysis and Applications* 2011 (2011).

[12] **Choi, W.-H. and Park, C.-y.** *A Simple Line Coupler with Adaptive Impedance Matching for Power Line Communication.* In: *IEEE International Symposium on Power Line Communications and Its Applications (ISPLC).* 2007, pp. 187–191.

[13] **Corripio, F. J. C., Río, L. D. del, and Muňoz, J. E.** *A Time Variant Model for Indoor Power-Line Channels.* In: *Proceedings of the International Symposium on Power Line Communications.* 2001, pp. 85–90.

[14] **Corripio, F. J. C. et al.** *Analysis of the Cyclic Short-Term Variation of Indoor Power Line Channels.* In: *IEEE Journal on Selected Areas in Communications* 24.7 (2006), pp. 1327–1338.

[15] **Corripio, F. J. C. et al.** *Time-Varying Channel Emulator for Indoor Power Line Communications.* In: *Proceedings of the 51th Annual IEEE Global Telecommunications Conference (GLOBECOM).* 2008, pp. 1–5.

[16] **Cortés, J. A., Diez Luis andCorripio, F. J. C., and Sánchez-Martínez, J. J.** *Analysis of the indoor broadband power-line noise scenario.* In: *IEEE Transactions on Electromagnetic Compatibility* 52.4 (2010), pp. 849–858.

[17] **Cortés, J. A. et al.** *Analysis of the Periodic Impulsive Noise Asynchronous with the Mains in Indoor PLC Channels.* In: *IEEE International Symposium on Power Line Communications and Its Applications (ISPLC).* IEEE. 2009, pp. 26–30.

[18] **Daher, A. et al.** *Overlap-Save and Overlap-Add Filters: Optimal Design and Comparison.* In: *IEEE Transactions on Signal Processing* 58.6 (2010), pp. 3066–3075.

[19] **Dorf, R. C. and Svoboda, J. A.** *Introduction to Electric Circuits.* John Wiley & Sons, 2010.

[20] **Dostert, K.** *Powerline communications.* Prentice Hall PTR, 2001.

[21] **Dostert, K., Sigle, M., and Liu, W.** *Narrowband Characterisation in an Office Environment.* In: *MIMO Power Line Communications: Narrow and Broadband Standards, EMC, and Advanced Processing.* CRC Press, 2014, pp. 39–68.

[22] **Eldar, Y. C. and Kutyniok, G.** *Compressed Sensing: Theory and Applications.* Cambridge University Press, 2012.

[23] **Elgard, O.** *Electric Energy Systems Theory*. 1982.

[24] **Esmailian, T., Kschischang, F. R., and Glenn Gulak, P.** *In-Building Power Lines as High-Speed Communication Channels: Channel Characterization and a Test Channel Ensemble*. In: *International Journal of Communication Systems* 16.5 (2003), pp. 381–400.

[25] **Feng, M. et al.** *Artifact Removal for Intracranial Pressure Monitoring Signals: a Robust Solution with Signal Decomposition*. In: *Annual International Conference of the IEEE Engineering in Medicine and Biology Society (EMBC)*. IEEE. 2011, pp. 797–801.

[26] **Ferreira, H. C. et al.** *Power Line Communications: Theory and Applications for Narrowband and Broadband Communications over Power Lines*. John Wiley & Sons, 2011.

[27] **Galli, S.** *A Simplified Model for the Indoor Power Line Channel*. In: *IEEE International Symposium on Power Line Communications and Its Applications (ISPLC)*. IEEE. 2009, pp. 13–19.

[28] **Galli, S. and Banwell, T.** *A Novel Approach to the Modeling of the Indoor Power Line Channel - Part II: Transfer Function and Its Properties*. In: *IEEE Transactions on Power Delivery* 20.3 (2005), pp. 1869–1878.

[29] **Galli, S. and Banwell, T. C.** *A Deterministic Frequency-Domain Model for the Indoor Power Line Transfer Function*. In: *IEEE Journal on Selected Areas in Communications* 24.7 (2006), pp. 1304–1316.

[30] **Galli, S., Scaglione, A., and Wang, Z.** *For the grid and through the grid: The role of power line communications in the smart grid*. In: *Proceedings of the IEEE* 99.6 (2011), pp. 998–1027.

[31] **Gardner, W.** *Introduction to Random Processes with Applications to Signals and Systems*. In: (1985).

[32] **Gardner, W. et al.** *Cyclic Wiener Filtering: Theory and Method*. In: *IEEE Transactions on Communications* 41.1 (1993), pp. 151–163.

[33] **Gardner, W. A.** *The Spectral Correlation Theory of Cyclostationary Time-Series*. In: *Signal Processing* 11.1 (1986), pp. 13–36.

[34] **Gardner, W. A., Napolitano, A., and Paura, L.** *Cyclostationarity: Half a Century of Research*. In: *Signal Processing* 86.4 (2006), pp. 639–697.

[35] **Ghosh, M.** *Analysis of the Effect of Impulse Noise on Multicarrier and Single Carrier QAM Systems*. In: *IEEE Transactions on Communications* 44.2 (1996), pp. 145–147.

[36] **Götz, M.** *Mikroelektronische, echtzeitfähige Emulation von Powerline-Kommunikationskanälen*. Berlin: Mensch & Buch Verlag, 2004.

[37] **Götz, M., Rapp, M., and Dostert, K.** *Power Line Channel Characteristics and Their Effect on Communication System Design*. In: *Communications Magazine* 42.4 (2004), pp. 78–86.

[38] **Gröchenig, K.** *Foundations of Time-Frequency Analysis*. Springer Science & Business Media, 2013.

[39] **Hashmat, R. et al.** *Measurement and Analysis of Inhome MIMO PLC Channel Noise*. In: *Workshop on Power Line Communications (WSPLC)*. 2010, pp. 56–60.

[40] **Hashmat, R. et al.** *A Time-Domain Model of Background Noise for In-Home MIMO PLC Networks*. In: *IEEE Transactions on Power Delivery* 27.4 (2012), pp. 2082–2089.

[41] **Hashmat, R. et al.** *Analysis and Modeling of Background Noise for Inhome MIMO PLC Channels*. In: *IEEE International Symposium on Power Line Communications and Its Applications (ISPLC)*. IEEE. 2012, pp. 316–321.

[42] **Hooijen, O. G.** *A Channel Model for the Residential Power Circuit Used as a Digital Communications Medium*. In: *IEEE Transactions on Electromagnetic Compatibility* 40.4 (1998), pp. 331–336.

[43] **Huang, N. E. et al.** *The Empirical Mode Decomposition and the Hilbert Spectrum for Nonlinear and Non-Stationary Time Series Analysis*. In: *Proceedings of the Royal Society of London A: Mathematical, Physical and Engineering Sciences*. Vol. 454. 1971. The Royal Society. 1998, pp. 903–995.

[44] **Katayama, M., Yamazato, T., and Okada, H.** *A Mathematical Model of Narrowband Power-Line Noise Based on Measurements*. In: *IEEE International Symposium on Power Line Communications and its Applications (ISPLC)*. 2005.

[45] **Katayama, M., Yamazato, T., and Okada, H.** *A Mathematical Model of Noise in Narrowband Power Line Communication Systems*. In: *Journal on Selected Areas in Communications* 24.7 (2006), pp. 1267–1276.

[46] **Katayama, M. et al.** *Modeling of Cyclostationary and Frequency Dependent Power-Line Channels for Communications.* In: *IEEE International Symposium on Power Line Communication and its Applications (ISPLC.* 2000, pp. 123–130.

[47] **Khintchine, A.** *Korrelationstheorie der stationären stochastischen Prozesse.* In: *Mathematische Annalen* 109.1 (1934), pp. 604–615.

[48] **Kim, I. H. et al.** *Channel Measurements and Field Tests of Narrowband Power Line Communication over Korean Underground LV Power Lines.* In: *IEEE International Symposium on Power Line Communications and its Applications (ISPLC).* IEEE. 2014, pp. 132–137.

[49] **Koralov, L. and Sinai, Y. G.** *Theory of Probability and Random Processes.* Springer, 2007.

[50] **Langfeld, P.** *OFDM für die schnelle Powerline Communication.* Berlin: Mensch & Buch, 2002.

[51] **Li, Q., She, J., and Feng, Z.** *Adaptive Impedance Matching in Power Line Communication.* In: *IEEE International Conference on Microwave and Millimeter Wave Technology (ICMMT).* IEEE. 2004, pp. 887–890.

[52] **Lin, J. and Evans, B. L.** *Cyclostationary Noise Mitigation in Narrowband Powerline Communications.* In: *Asia-Pacific Signal & Information Processing Association Annual Summit and Conference (APSIPA ASC).* IEEE. 2012, pp. 1–4.

[53] **Liu, S. et al.** *Empirical mode decomposition applied to tissue artifact removal from respiratory signal.* In: *Engineering in Medicine and Biology Society, 2008. EMBS 2008. 30th Annual International Conference of the IEEE.* IEEE. 2008, pp. 3624–3627.

[54] **Liu, W., Bumiller, G., and Gao, H.** *On (Power-) Line Defined PLC System.* In: *IEEE International Symposium on Power Line Communications and its Applications.* IEEE. 2014, pp. 81–86.

[55] **Liu, W.** *Emulation of Narrowband Powerline Data Transmission Channels and Evaluation of PLC Systems.* Vol. 7. KIT Scientific Publishing, 2013.

[56] **Liu, W., Li, C., and Dostert, K.** *Emulation of AWGN for Noise Margin Test of Powerline Communication Systems.* In: *IEEE International Symposium on Power Line Communications and Its Applications (ISPLC).* IEEE. 2011, pp. 225–230.

[57] **Liu, W., Sigle, M., and Dostert, K.** *Advanced Emulation of Channel Transfer Functions for Performance Evaluation of Powerline Modems.* In: *IEEE International Symposium on Power Line Communications and Its Applications (ISPLC).* IEEE. 2011, pp. 446–451.

[58] **Liu, W., Sigle, M., and Dostert, K.** *Channel Characterization and System Verification for Narrowband Power Line Communication in Smart Grid Applications.* In: *IEEE Communications Magazine* 49.12 (2011), pp. 28–35.

[59] **MacQueen, J. et al.** *Some Methods for Classification and Analysis of Multivariate Observations.* In: *Proceedings of the Fifth Berkeley Symposium on Mathematical Statistics and Probability.* Vol. 1. 14. Oakland, CA, USA. 1967, pp. 281–297.

[60] **Mehr, A. S. and Chen, T.** *Properties of Linear Switching Time-Varying Discrete-Time Systems with Applications.* In: *Systems & Control Letters* 39.4 (2000), pp. 229–235.

[61] **Meng, H. et al.** *Modeling of Transfer Characteristics for the Broadband Power Line Communication Channel.* In: *IEEE Transactions on Power Delivery* 19.3 (2004), pp. 1057–1064.

[62] **Middleton, D.** *Procedures for Determining the Parameters of the First-Order Canonical Models of Class A and Class B Electromagnetic Interference.* In: *IEEE Transactions on Electromagnetic Compatibility* 3 (1979), pp. 190–208.

[63] **Mutagi, R.** *Pseudo Noise Sequences for Engineers.* In: *Electronics & Communication Engineering Journal* 8.2 (1996), pp. 79–87.

[64] **Nason, G. P., Von Sachs, R., and Kroisandt, G.** *Wavelet Processes and Adaptive Estimation of the Evolutionary Wavelet Spectrum.* In: *Journal of the Royal Statistical Society: Series B (Statistical Methodology)* 62.2 (2000), pp. 271–292.

[65] **Nassar, M. et al.** *Cyclostationary Noise Modeling in Narrowband Powerline Communication for Smart Grid Applications.* In: *IEEE International Conference on Acoustics, Speech and Signal Processing (ICASSP).* IEEE. 2012, pp. 3089–3092.

[66] **Nassar, M. et al.** *Local Utility Power Line Communications in the 3–500 kHz Band: Channel Impairments, Noise, and Standards.* In: *IEEE Signal Processing Magazine* 29.5 (2012), pp. 116–127.

[67] **Nieman, K. F. et al.** *Cyclic Spectral Analysis of Power Line Noise in the 3–200 kHz Band.* In: *IEEE International Symposium on Power Line Communications and Its Applications (ISPLC).* IEEE. 2013, pp. 315–320.

[68] **Ohno, O. et al.** *A Simple Model of Cyclostationary Power-Line Noise for Communication Systems.* In: *IEEE International Symposium on Power Line Communications and its Applications (ISPLC).* 1998, pp. 115–122.

[69] **Opalko, O., Sigle, M., and Dostert, K.** *Detection of AM Broadcasting Stations on Low Voltage Power Lines.* In: *IEEE International Symposium on Power Line Communications and its Applications (ISPLC).* IEEE, 2014, pp. 87–92.

[70] **Oppenheim, A. V., Schafer, R. W., Buck, J. R., et al.** *Discrete-Time Signal Processing.* Prentice-Hall Englewood Cliffs, 1989.

[71] **Pagani, P. et al.** *European MIMO PLC Field Measurements: Noise Analysis.* In: *IEEE International Symposium on Power Line Communications and Its Applications (ISPLC).* IEEE. 2012, pp. 310–315.

[72] **Philipps, H.** *Modelling of Powerline Communication Channels.* In: *IEEE International Symposium on Power Line Communications and Its Applications (ISPLC).* 1999, pp. 14–21.

[73] **Pighi, R. et al.** *Fundamental Performance Limits of Communications Systems Impaired by Impulse Noise.* In: *IEEE Transactions on Communications* 57.1 (2009), pp. 171–182.

[74] **Proakis, J. G.** *Digital Communications.* McGraw-Hill, New York, 2008.

[75] **Puente León, F., Kiencke, U., and Jäkel, H.** *Signale und Systeme.* Oldenbourg Verlag, München, 2010.

[76] **Rende, D. et al.** *Noise Correlation and Its Effect on Capacity of Inhome MIMO Power Line Channels.* In: *IEEE International Symposium on Power Line Communications and Its Applications (ISPLC).* IEEE. 2011, pp. 60–65.

[77] **Sanabria, S. D. A., Bonilla, V. R. P., and Becerra Tobar, J. M. D.** *Test Methodology for Impedance Characterization of a Low Voltage Power Line Channel for Broadband Communication Applications.* In: *Colombian Conference on Communications and Computing (COLCOM).* IEEE. 2014, pp. 1–6.

[78] **Sancha, S. et al.** *A Channel Simulator for Indoor Power-Line Communications.* In: *IEEE International Symposium on Power Line Communications and Its Applications (ISPLC).* IEEE. 2007, pp. 104–109.

[79] **Schneider, D. et al.** *European MIMO PLC Field Measurements: Channel Analysis*. In: *IEEE International Symposium on Power Line Communications and Its Applications (ISPLC)*. IEEE. 2012, pp. 304–309.

[80] **Schwager, A.** *Powerline Communications: Significant Technologies to Become Ready for Integration*. PhD thesis. Universität Duisburg-Essen, Fakultät für Ingenieurwissenschaften» Elektrotechnik und Informationstechnik» Energietransport und-speicherung, 2010.

[81] **Schwager, A. et al.** *MIMO PLC: Theory, Measurements and System Setup*. In: *IEEE International Symposium on Power Line Communications and Its Applications (ISPLC)*. IEEE. 2011, pp. 48–53.

[82] **Sigle, M., Liu, W., and Dostert, K.** *On the Impedance of the Low-Voltage Distribution Grid at Frequencies Up to 500 kHz*. In: *IEEE International Symposium on Power Line Communications and Its Applications (ISPLC)*. IEEE. 2012, pp. 30–34.

[83] **Silverman, R.** *Locally Stationary Random Processes*. In: *IRE Transactions on Information Theory* 3.3 (1957), pp. 182–187.

[84] **Stadelmeier, L. et al.** *MIMO for In-Home Power Line Communications*. In: *International ITG Conference on Source and Channel Coding (SCC)*. VDE. 2008, pp. 1–6.

[85] **Sugiura, Y., Yamazato, T., and Katayama, M.** *Measurement of Narrowband Channel Characteristics in Single-Phase Three-Wire Indoor Power-Line Channels*. In: *IEEE International Symposium on Power Line Communications and Its Applications (ISPLC)*. IEEE. 2008, pp. 18–23.

[86] **Sun, Y. and Amaratunga, G.** *High-Current Adaptive Impedance Matching in Narrowband Power-Line Communication Systems*. In: *IEEE International Symposium on Power Line Communications and Its Applications (ISPLC)*. IEEE. 2011, pp. 329–334.

[87] **Sung, T.-E., Scaglione, A., and Galli, S.** *Time-Varying Power Line Block Transmission Models Over Doubly Selective Channels*. In: *IEEE International Symposium on Power Line Communications and Its Applications (ISPLC)*. IEEE. 2008, pp. 193–198.

[88] **Tlich, M. et al.** *Indoor Power-Line Communications Channel Characterization Up to 100 MHz — Part I: One-Parameter Deterministic Model*. In: *IEEE Transactions on Power Delivery* 23.3 (2008), pp. 1392–1401.

[89] **Tomasoni, A., Riva, R., and Bellini, S.** *Spatial Correlation Analysis and Model for In-Home MIMO Power Line Channels*. In: *IEEE International Symposium on Power Line Communications and Its Applications (ISPLC)*. IEEE. 2012, pp. 286–291.

[90] **Tonello, A., Versolatto, F., and Pittolo, A.** *In-Home Power Line Communication Channel: Statistical Characterization*. In: *IEEE Transactions on Communications* 62.6 (2014), pp. 2096–2106.

[91] **Van Rensburg, P. J.** *Effective coupling for power-line communications*. PhD thesis. University of Johannesburg, 2008.

[92] **Weling, N.** *Flexible FPGA-based Powerline Channel Emulator for Testing MIMO PLC, Neighborhood Networks, Hidden Node or VDSL Coexistence Scenarios*. In: *IEEE International Symposium on Power Line Communications and Its Applications (ISPLC)*. IEEE. 2011, pp. 12–17.

[93] **Weling, N., Engelen, A., and Thiel, S.** *Broadband MIMO Powerline Channel Emulator*. In: *IEEE International Symposium on Power Line Communications and its Applications (ISPLC)*. IEEE. 2014, pp. 105–110.

[94] **Weling, N., Engelen, A., and Thiel, S.** *Broadband MIMO Powerline Channel Emulator Verifb01cation and Testing Results*. In: *IEEE International Symposium on Power Line Communications and its Applications (ISPLC)*. IEEE. 2015, pp. 59–64.

[95] **Zhao, Z. and Ma, C.** *A Novel Cancellation Method of Powerline Interference in ECG Signal Based on EMD and Adaptive Filter*. In: *IEEE International Conference on Communication Technology (ICCT)*. IEEE. 2008, pp. 517–520.

[96] **Zimmerman, M. and Dostert, K.** *A Multi-Path Signal Propagation Model for the Powerline Channel in the High Frequency Range*. In: *IEEE International Symposium on Power Line Communications and Its Applications (ISPLC)*. 1999, pp. 45–51.

[97] **Zimmermann, M.** *Energieverteilnetze als Zugangsmedium für Telekommunikationsdienste*. PhD thesis. 2000.

[98] **Zimmermann, M. and Dostert, K.** *An Analysis of the Broadband Noise Scenario in Powerline Networks*. In: *IEEE International Symposium on Powerline Communications and its Applications (ISPLC)*. 2000, pp. 5–7.

[99] **Zimmermann, M. and Dostert, K.** *A Multipath Model for the Powerline Channel*. In: *IEEE Transactions on Communications* 50.4 (2002), pp. 553–559.

[100] **Zimmermann, M. and Dostert, K.** *Analysis and Modeling of Impulsive Noise in Broad-Band Powerline Communications*. In: *IEEE Transactions on Electromagnetic Compatibility* 44.1 (2002), pp. 249–258.

List of Publications

[101] **Han, B., Kaiser, C., and Dostert, K.** *A Novel Approach of Canceling Cyclostationary Noise in Low-Voltage Power Line Communications*. In: *IEEE International Conference on Communications (ICC)*. IEEE. 2015.

[102] **Han, B. et al.** *An Online Approach for Intracranial Pressure Forecasting Based on Signal Decomposition and Robust Statistics*. In: *IEEE International Conference on Acoustics, Speech and Signal Processing (ICASSP)*. IEEE. 2013, pp. 6239–6243.

[103] **Otterbach, N. et al.** *Software Defined Radio for Power Line Communication Research and Development*. In: *Software Radio Implementation Forum (SRIF)*. ACM. 2015.

List of Supervised Theses

[104] **Kaiser, C.** *Analyse und Filterung des periodischen, additiven und impulsiven Rauschens in Powerline-Kanälen*. Masterarbeit. IIIT, 2014.

Forschungsberichte aus der Industriellen Informationstechnik (ISSN 2190-6629)

Institut für Industrielle Informationstechnik
Karlsruher Institut für Technologie (KIT)

Hrsg.: Prof. Dr.-Ing. Fernando Puente León, Prof. Dr.-Ing. habil. Klaus Dostert

Die Bände sind unter www.ksp.kit.edu als PDF frei verfügbar oder als Druckausgabe bestellbar.

Band 1 Pérez Grassi, Ana
Variable illumination and invariant features for detecting and classifying varnish defects. (2010)
ISBN 978-3-86644-537-6

Band 2 Christ, Konrad
Kalibrierung von Magnet-Injektoren für Benzin-Direkteinspritzsysteme mittels Körperschall. (2011)
ISBN 978-3-86644-718-9

Band 3 Sandmair, Andreas
Konzepte zur Trennung von Sprachsignalen in unterbestimmten Szenarien. (2011)
ISBN 978-3-86644-744-8

Band 4 Bauer, Michael
Vergleich von Mehrträger-Übertragungsverfahren und Entwurfskriterien für neuartige Powerline-Kommunikationssysteme zur Realisierung von Smart Grids. (2012)
ISBN 978-3-86644-779-0

Band 5 Kruse, Marco
Mehrobjekt-Zustandsschätzung mit verteilten Sensorträgern am Beispiel der Umfeldwahrnehmung im Straßenverkehr (2013)
ISBN 978-3-86644-982-4

Band 6 Dudeck, Sven
Kamerabasierte In-situ-Überwachung gepulster Laserschweißprozesse (2013)
ISBN 978-3-7315-0019-3

Band 7 Liu, Wenqing
Emulation of Narrowband Powerline Data Transmission Channels and Evaluation of PLC Systems (2013)
ISBN 978-3-7315-0071-1

Forschungsberichte aus der Industriellen Informationstechnik (ISSN 2190-6629)
Institut für Industrielle Informationstechnik | Karlsruher Institut für Technologie (KIT)

Hrsg.: Prof. Dr.-Ing. Fernando Puente León, Prof. Dr.-Ing. habil. Klaus Dostert

Band 8 Otto, Carola
Fusion of Data from Heterogeneous Sensors with Distributed Fields of View and Situation Evaluation for Advanced Driver Assistance Systems. (2013)
ISBN 978-3-7315-0073-5

Band 9 Wang, Limeng
Image Analysis and Evaluation of Cylinder Bore Surfaces in Micrographs. (2014)
ISBN 978-3-7315-0239-5

Band 10 Michelsburg, Matthias
Materialklassifikation in optischen Inspektionssystemen mithilfe hyperspektraler Daten. (2014)
ISBN 978-3-7315-0273-9

Band 11 Pallauf, Johannes
Objektsensitive Verfolgung und Klassifikation von Fußgängern mit verteilten Multi-Sensor-Trägern. (2016)
ISBN 978-3-7315-0529-7

Band 12 Sigle, Martin
Robuste Schmalband-Powerline-Kommunikation für Niederspannungsverteilernetze. (2016)
ISBN 978-3-7315-0539-6

Band 13 Opalko, Oliver
Powerline-Kommunikation für Batteriemanagement-Systeme in Elektro- und Hybridfahrzeugen. (2017)
ISBN 978-3-7315-0647-8

Band 14 Han, Bin
Characterization and Emulation of Low-Voltage Power Line Channels for Narrowband and Broadband Communication. (2017)
ISBN 978-3-7315-0654-6